# Fishers at Work, Workers at Sea

## *A Puerto Rican Journey through Labor and Refuge*

# Fishers at Work, Workers at Sea

*A Puerto Rican Journey through Labor and Refuge*

DAVID GRIFFITH AND
MANUEL VALDÉS PIZZINI

TEMPLE UNIVERSITY PRESS
Philadelphia

Temple University Press, Philadelphia 19122
Copyright © 2002 by Temple University
All rights reserved
Published 2002
Printed in the United States of America

♾ The paper used in this publication meets the requirements of the
American National Standard for Information Sciences—Permanence
of Paper for Printed Library Materials, ANSI Z39.48-1984

Library of Congress Cataloging-in-Publication Data

Griffith, David Craig, 1951–
   Fishers at work, workers at sea : a Puerto Rican journey through labor
and refuge / David Griffith and Manuel Valdés Pizzini.
      p.   cm.
   Includes bibliographical references and index.
   ISBN 1-56639-910-6 (cloth : alk. paper) — ISBN 1-56639-911-4 (paper-
back : alk. paper)
      1. Fishers—Puerto Rico.   2. Fishers—Puerto Rico—Interviews.
   3. Fisheries—Puerto Rico.   4. Wages—Fishers—Puerto Rico.
   5. Household surveys—Puerto Rico.   6. Puerto Ricans—Florida.
   7. Puerto Ricans—New York (State)—New York.   8. Migrant labor—
   Puerto Rico.   9. Migrant labor—United States.   I. Valdés Pizzini,
   Manuel, 1954– .   II. Title.

   HD8039.F652 P94   2002
   331.7'6392'097295–dc21                                      2001027640

Much of this research was funded by the National Science Foundation
under a project entitled Wage Labor and Small-Scale Fishing in Puerto
Rico (BNS-8718670).

Excerpt (p. 131) from A WALK WITH TOM JEFFERSON by Philip Levine,
copyright © 1988 by Philip Levine. Used by permission of Alfred A.
Knopf, a division of Random House, Inc.

Excerpt (pp. 246–47) from "Santarém" from THE COMPLETE POEMS
1927–1979 by Elizabeth Bishop. Copyright © 1979, 1983 by Alice Helen
Methfessel. Reprinted by permission of Farrar, Straus and Giroux, LLC.

*For William Roseberry*
y paz para Vieques

# Contents

*Photographs follow page 120*

# Preface

PEOPLE AROUND THE WORLD are trying to maintain their hold on cultural and natural resources that provide them with primary and supplemental incomes that many of us would consider quite meager. Much of this income enters family settings in the form of subsistence; yet much also enters the store of family security in the form of feelings of identity, belonging, and self-worth. Those who fish, hunt, farm, gather, raise animals, or rely on other natural resources for some or most of their survival often report that they gain much satisfaction from these activities beyond the more tangible benefits of the cash, food, or other materials that they provide. Most important, these activities often lie at the core of small-scale producers' identities: People strive to maintain and reproduce these ways of life explicitly because, by doing so, they maintain and reproduce their cultural heritage.

In this book, investigating wage-labor activity among small-scale fishers in Puerto Rico, we seek to elucidate how the incorporation of independent producers into labor and commodity markets affects their economic activities and their sense of self. Among the issues addressed are the ways that different levels of wage labor among fishers affect their resource exploitation and marine ecosystems; the allocation of household tasks and relations among fishing households; political activity; investment; and fishers' cognitive representations of fishing, work, and labor. The techniques we employed included eliciting fishers' life histories (our primary method), repeated observations of fishing and domestic producer activities, structured tasks (the pile sorting of fishing activities and occupations), and secondary source data collection. All quotations that are not otherwise cited are taken from the authors' field notes or interviews.

Of our 102 sample households, fewer than 20 percent had never contributed a member to wage-labor markets. Variations in levels of wage-labor activity among the households indicate at least three paths of change: increased dependence on wagework, decreased dependence on wagework, and maintaining a balance between wagework and fishing. Each path implies different relations among fishing households, different levels of political activity in fishing associations, the use of different fishing techniques, the pursuit of different species, and the exploitation of marine environments. High levels of wage-labor activity, for example, seem to be associated with types of gear and fishing strategies that do not require extensive reliance on interhousehold relationships, whereas reduced dependence on wage labor usually involves the development of new relationships among households or the utilization of existing relationships to intensify fishing activities.

Most of the research and the original ideas for this book were funded by the National Science Foundation (NSF), which was kind enough to support a project entitled Wage Labor and Small-Scale Fishing in Puerto Rico (BNS-8718670). Any errors in the text are, of course, the responsibility of the authors. Along with the authors of this book, friend and colleague Jeff Johnson participated in this project, offering many methodological suggestions. In the field, we were fortunate to have the help of a remarkable group of assistants (all students) who immersed themselves in the interview process with zeal, professionalism, and commitment. Their contribution to this book is duly recognized here: We thank Maria Cruz Torres, Manuel Huerta, Carmen Gloria Castro, Marisol Camacho, Zoraida Santiago, Roxanna Fernández, Yasmín Detrés, Carmen Milagros Márquez, Juan Vera, and Nitza Seguí. The quality of their work on this project is also reflected in the success of their careers in anthropology (Cruz Torres and Santiago), in community work (Fernández and Seguí), in public service (Castro), in oceanography (Detrés), in teaching and research (Huerta and Camacho), and in underwater archaeology (Márquez and Vera).

Additional support for some of the research that informed this work was provided by the National Oceanographic and Atmospheric Administration (NOAA). The Southeast Regional Office of the National Marine Fisheries Service (NMFS) funded our inventory of fishing infrastructure around Puerto Rico and the U.S. Virgin Islands, a study that we conducted with the aid and expertise of Ruperto Chaparro (University of Puerto Rico Sea Grant College Program) and Jaime Gutiérrez Sánchez (University of Puerto Rico at Mayagüez). Research activities on fisher associations and the fishing community of Puerto Real was largely funded by the University of Puerto Rico Sea Grant College Program (Grant SE/D-20-1), in a project designed and supervised by Jaime Gutiérrez Sánchez and Bonnie McCay. Our studies of the life histories of gill net and trammel net fishers were made possible by a grant from the NMFS, through the Saltonstall-Kennedy Funds. We thank Alejandro Acosta and Mervin Ruíz for their collaboration on many aspects of that project.

We have both benefited a great deal from long association with the Sea Grant College Programs of the University of Puerto Rico and the University of North Carolina. Drs. Manuel Hernández Avila, Shirley Fiske, James Murray, B. J. Copeland, and Ron Hodson all contributed much in the way of general research support over the past two decades, both before and after most of the research that produced this volume was conducted. Their vision and the support they have given to social science research should be commended, and it is warmly appreciated by the authors. Time was allotted to Valdés Pizzini for field activities thanks to the Marine Advisory Services Project of the University of Puerto Rico Sea Grant College Program.

We also thank our many other colleagues at East Carolina University and the University of Puerto Rico. First, we express our deepest gratitude to Bill Queen, director of the Institute for Coastal and Marine Resources. Among the many East Carolina University colleagues who contributed to this project are Lori King, Director of the Coastal Resources Management Program; Tom Feldbush, Vice Chancellor for Academic Affairs; and Keats Sparrow, Dean of Arts and Sciences. We greatly appreciate the help of Gutiérrez Sánchez at the University of Puerto Rico, who gave Valdés Pizzini the opportunity to work with him on various projects concerning the local fisheries, projects that eventually led to our study. Gutiérrez Sánchez also had the vision to create a center for applied social research; he entrusted Valdés Pizzini with the mission of developing the Centro de Investigación Social Aplicada, under whose auspices portions of the research and the writing for this book were completed. In addition, the kindness and support of Don Jaime, Chair of the Social Science Department, are greatly appreciated.

Throughout this rather long journey, many good friends and colleagues took some of their precious time to lend us a hand with this project in a wide variety of ways. We thank Mario Nuñez, Ricardo Pérez, Carlos Buitrago Ortiz, Miguel Rolón, Joseph Kimmel, Graciela García Moliner, Bonnie McCay, Michael Orbach, and John Poggie. Our special thanks go to E. Paul Durrenberger and Robert Lee Maril for their editorial comments and encouragement.

Puerto Rican fishers and their families deserve our deepest gratitude. They opened their doors and allowed us into their homes, yards, piers, and boats to ask them difficult questions about their lives and trajectories and to participate in their ritual, political, and productive activities. Their collaboration was crucial, but naturally it came with a degree of the contentious stance that characterizes their praxis. Our steps were scrutinized. We were interrogated, tested, observed (almost ethnographically), dodged (we still remember him), questioned, and even frisked for weapons and police identification on an isolated hill in Vieques. But by a great margin, we benefited from their magnanimous hospitality, even—several times throughout the period of this project—in individual homes, which provided us with hours of interviews and conversations that helped us gain some comprehension of their complex lives and labor trajectories in a globalized world that they fully understand. We acknowledge their collective contribution to this book. And, on a personal note, we thank a number of these people, who, for many reasons, we remember with special warmth and appreciation: Roberto Franqui, Sr.; Roberto Franqui, Jr.; Luís Ortiz; Elton Ortiz; Yolanda Rodríguez; Manuel Ventura; Rodrigo Irizarry; José Beza; and Hiram (Froilan) López.

The woes of the Puerto Rican fisher are many and varied, as this book illustrates. However, the fishers of Vieques have endured the difficult process of contending with the U.S. Navy in the dispute over their land (which is used as a target range) and their right to have access to the sea. They have our admiration and our best hopes of *paz para Vieques* (peace for Vieques).

In my case (Valdés Pizzini), this project came at a critical professional juncture, which increased the pressure on family members. I was extremely fortunate to have the support of my family. My wife, Zaida, and my daughters, Ana Krystalliá and Carmen Margarita, endured the process with warmth, love, and an understanding much greater than I deserved. Griffith thanks his wife, Nancy, and his daughters, Emily and Brook.

Finally, we must add a sad note. During the final days of manuscript preparation, William Roseberry, a dear colleague and deeply insightful anthropologist, died at the age of fifty. In the late 1970s, Griffith had the excellent fortune to study with Bill as one of his graduate students at the University of Iowa. Roseberry's insightful reflections on the 1956 work by Julian H. Steward and colleagues, *The People of Puerto Rico*, have always been critical in Valdés Pizzini's work. Moreover, Roseberry's 1976 *American Anthropologist* article on rent and differentiation among peasants helped Valdés Pizzini turn a corner in the writing of his dissertation on capital formation among the fishers of Puerto Real. We discussed and drew upon Roseberry's work throughout the development of this book, as far back as the original research concepts in the proposal that we submitted to the NSF. In fact, Roseberry commented on early drafts of this proposal and offered suggestions for its improvement. We would like to believe that his genius and particularly his insistence on the examination of the intersections of global and local histories shine through much of this book; certainly his love of history has influenced its development and our understanding of the complex processes that face Latin America and Puerto Rico today. Roseberry's essay "Americanization in the Americas," which directly influenced this work, ends with the following observation:

> Modernization, Westernization, and Americanization can imply linear processes connecting polar opposites. . . . Perhaps the operative label should not be Americanization but Puerto-Ricanization, Mexicanization, Peruvianization, and so on. We are still dealing with "ization" words, but words that direct us to specific historical processes. There is always a danger that the latter set of words can be placed at a polar extreme from "Americanization" as part of a romantic search for cultural authenticity, an artificial separation of the history of capitalism

(or, in this case, the history of U.S. expansion) from a society's "own" history. This too would be a mistake. The understanding of any of these processes would direct us to powerful external forces, especially, in this century, the United States. By placing an emphasis on particular national experiences, however, we can see that these forces are inserted in particular contexts of power, each of which represents particular internalizations of the external. (1989:120–21)

We hope that, in the pages that follow, we have taken Roseberry's advice with nearly as much rigor and understanding as he would have brought to a work like this.

# 1     Divided Selves

## *Domestic Production and Wage Labor in Puerto Rico and Anthropology*

THE FIELD BOSS speaks to the crew in the broken Spanish he picked up from the Mexican crews before they left for
the Blue Ridge Mountains to shape Christmas trees. It is the
tail end of a long Indian summer. Back home in Puerto Rico,
where Ángel and Miguel have not set foot since June, the
threat of hurricanes is passing for another year. Standing in
the half circle around the field boss, listening to him explain
the tasks of harvest as though teaching dogs how to dig for
gophers, Ángel and Miguel do not let on that they speak
fairly good English. Not only is it fun to watch him form his
words so slowly and carefully; it is also possible to make
your job easier by pretending not to understand. *"Ponen los
repollos aquí,"* says the field boss, pointing to the big
wooden crates in the shade of the packing shed. Ángel tries
to look confused, fearing that he looks amused instead. If he
were to catch the eye of Miguel or any one of the other crew
members, he would not be able to keep from laughing.

At least this crew boss makes an attempt to speak Spanish.
Ángel has known field bosses who believed that anyone anywhere could understand English, as long as it was spoken
slowly enough. Ángel and the others especially like to watch
the field boss demonstrate how and where the heads of cabbage
go. They want to see the field boss walk the crew through all
the harvest motions. They want to see him do it himself. Only
by performing the work himself can the boss appreciate the
hardships of working in the U.S. *fincas.* Only by doing it himself can he feel the pains shooting sideways across his lower
back after stooping and standing, stooping and standing, for ten
hours a day; or the sun burning his neck; or the pesticide

1

residues stinging his eyes and nose, the toxic dust itching his skin. Ángel believes the field bosses should know: The more a *jefe* experiences the work of those he directs, the closer he comes to passing along the work's tradition, its reason, its necessities of skill and of craft, just as Ángel's father, a fisher, passed his knowledge of fishing to his son. Ángel's father's lessons were deep and comprehensive, nothing like those that have ever come from the field boss, as interwoven with the life of his childhood home as his father's beach seine. He remembers his mother scaling and cleaning the catch for market, the fishers of the community organizing against the Department of Natural Resources (DNR) for the right to cut away mangroves to make a place to launch their fishing crafts into the sea, and his brother painting the hull of their fourteen-foot *yola*. These were as much the lessons of fishing as the way his father took him through the stages of making and setting fish traps, bending and soldering the rods, wiring on the chicken wire and weaving cut bamboo through it to maintain the integrity of the shape of the trap. While they worked, his father's game cocks crowed from their cages along the walls of the workshop. In order to outwit thieves, they had to set the traps in secret, without buoys, within sight of landmarks to mark their location. Remembering, he thinks of the names of fish and other marine animals they caught in the traps: *cabrilla, colirrubia, sama, pargo, chillo, mero,* and *langosta,* the reef and bottom fish that bring high prices in the seafood restaurants between Mayagüez and Puerto Real; or *chapín,* the little box-shaped fishes whose meats fill the *empanadillas* (seafood pastries) of La Parguera; or *pulpo,* the octopus that the women of Las Croabas boil, let cool, season with onions, oil, vinegar, and green olives, and sell in little plastic cups along the main roads to Fajardo on weekends. These are bottom-feeding species, the life of the reef. With other gear (*corrida, silga, cordeles, anzuelos*), they catch the fish that cruise near the surface during the winter months (*sierra, dorado, atún*), the fish that strike and fight with their long, sleek bodies, only to be sliced into steaks that the vendors and shopkeepers fry with sweet onions and garlic to sell in the streets of town.

It is not only watching the field boss take the crew through the motions of harvest that makes Ángel think of his father's lessons. The memory also comes from the distinctive smell of the ocean, perceptible even over the cabbage fields twelve miles inland from Cape May. It is a therapeutic smell, one that calls to mind the quiet, open stillness of the sea, free of the crack, aspirin, beer, and cheap overproof wine that are always central to life in the isolated farm labor camps. The smell calls to mind a fishing excursion achieved only after minute and detailed preparations, the close attention to every piece of gear so absorbing that the whole world seems to hold still to let you finish the meticulous tasks of rigging lines, tying feathers onto hooks in an overlapping, spiral arrangement, and straightening the wire mesh of the fish traps. Here the sea is doubly therapeutic, taking Ángel back to his work and homeland at the same time that it offers the steady breeze that keeps the air above the fields relatively free of pesticide fumes.

But the field boss has finished his demonstration. Stooping, Ángel loses the full effect of the breeze, smelling the chemical pungency that accounts for the headaches and bellyaches that plague the camp all through the harvest. In a smooth, rapid motion, he bends and cuts the cabbage and bends and cuts again and again, bagging the heads, moving with the crew down the row.

## Moving Between Domestic Production and Wage Labor

Like Ángel, who performs farmwork in southern New Jersey while he remains very much a part of his father's household and fishing operation, many people from poor households around the world combine wage labor with small-scale productive and reproductive activities that require small investments of capital and energy. A rural shopkeeper in Oaxaca, Mexico, travels to Guadalajara, Mexico, to lay bricks during a construction boom and returns to his shop when the building he helped construct is completed. A Jamaican youth works part time in a jewelry store and spends

his evenings picking pockets and selling ganja (the local mari-
juana) to British and American tourists and businesspeople stay-
ing in New Kingston. A Puerto Rican chambermaid in San Juan's
Condado district quits her job during the slow summer months
to sell earrings and necklaces that she makes out of tarpon scales,
tropical shells, and the spinal cartilage of juvenile sharks. Or a
Kikuyu peasant farmer spends twelve years of his life emptying
trash in a London bank building to accumulate enough wealth to
expand his cattle herd and bargain for a wife in Kenya.

Each of these examples entails a different amount and quality
of involvement in wage labor. Each influences social class formation differently. Each stimulates the development of new, creative relationships among businesses, workers and unions, households, governments, and the neighborhoods and communities
that supply workers to the formal labor force. For the rural shopkeeper, Guadalajara's booming economy provides a one-time,
short-term employment opportunity that involves internal migration, in which the contract between employer and employed
appears limited to the simple exchange of the shopkeeper's labor
power for the employer's wage. This image of equivalent
exchange—common under capitalism—disguises the systems of
authority and policy and the need they create that bring buyers
and sellers of labor together. The urban youth in Kingston combines wage labor and independent economic activity daily or at
least several times a week. He does not migrate; yet his work
requires passage from the formal to the informal economy: from
the safe, low-paying job in the jewelry store to the shadowy, dangerous, yet potentially high-return frontiers of crime and the
night. The artisan/chambermaid moves annually between a seasonal labor market and a cottage industry, much in the same way
that Ángel moves between farmwork in South Jersey and small-
scale fishing in Puerto Rico. Finally, the Kikuyu peasant, the jan-
itor/farmer, represents a type of adaptation common during the
1950s and 1960s, when Southern Europeans, Asians, Africans,
West Indians, and Latin Americans poured into Europe and North
America during the postwar economic expansion and then, late
in life, retired to their homelands (Basch, Glick-Schiller, and Szanton Blanc 1994; Brandes 1975; Gmelch 1987).

These examples cover but a fraction of the ground. As common methods of mixing wage labor with seemingly independent, autonomous economic activity, they include many of the arrangements that we encounter in the world today. Yet if any lesson emerges from this work, it is that combining wage labor with home production is a dynamic process, formulated, conditioned, and revised by the characters of both capital and domestic time and space. Such multiple livelihoods are not beyond the reach of explanation. Yet they demand that we draw on theoretical work on proletarians along with theoretical work on peasant, artisanal, and small-scale producers, negotiating among explanations even as the categories of proletarian, peasant, artisanal fisher, and others come under attack (Kearney 1996).

In this work, we look closely at the ways that Puerto Rican small-scale fishers combine fishing and wagework. Puerto Rican fishing families, like most families around the world, rarely rely on a single economic pursuit to survive. Instead, they combine fishing and fish vending with jobs in public service, agriculture, and industry and more informal enterprises such as guiding tourists through mangrove forests to bioluminescent bays or making crafts from shark cartilage, tarpon scales, and shells. In this book, we examine these multiple livelihoods up close. We consider what they mean for theories drawn from ecological and economic anthropology and for how we think about households of small producers everywhere. In the process, we discuss several concepts and assess them in light of the behaviors of Puerto Rican fishers. We consider the variety of ways that one key concept, proletarianization, or the process of *becoming* a wage laborer, is expressed in Puerto Rican fishing households and communities. Most commonly, we have found that these households and communities tend to be only *semi*proletarianized: partially engaged in wage labor and working-class behaviors and beliefs yet still strongly present in fishing. Semiproletarianization often involves the combination of formal and informal economic activities, with the former consisting of activities that governments and economists consider legitimate and that are taxed, regulated, and largely institutionalized. Working the disassembly line in a poultry plant, though usually unpleasant, is a formal economic activity. Informal economic activities consist of those that people use, even in the face of laws that prohibit them, to make ends meet. Taking fish out of season or selling octopus salad along the roadside without a license, though both may be pleasant activities, occur beyond the reach of the state and most economic analysis.

Because semiproletarianization almost inevitably involves households, neighborhoods, fishing associations, and coastal communities—areas where fishers and workers are reproduced and acquire a sense of humanity—we also consider relations between production and reproduction and the roles of these collectives in such relations. Particularly important are households, whose compositions and life cycles influence, quite strongly, the ways in which fishers combine fishing with jobs and other economic pursuits. Yet the collectives that we identify as households among Puerto Rican fishers differ from many common representations of households in the sociological and anthropological literature. We are indebted to several previous works on independent-producer households (see, for example, Chayanov 1966; Deere and de Janvry 1979; Durrenberger 1995; Gudeman and Rivera 1993; Roseberry 1976, 1989). But we have also drawn on literature on households that come from long traditions of emigration and return migration (see, for example, Basch, Glick-Schiller, and Szanton Blanc 1994; Duany 2000; Portes and Bach 1985).

## METHODOLOGICAL AND SAMPLING CONSIDERATIONS

The bulk of our work in this volume focuses on 102 fishing households in 21 coastal municipalities on the main island of Puerto Rico and the island municipalities of Culebra and Vieques. We sampled disproportionately more households from seven municipalities for the following reasons:

1. *Cabo Rojo,* on the Southwest Coast, includes Puerto Real, the port of Puerto Rico's largest fishery and the subject of Manuel Valdés Pizzini's doctoral dissertation (1985). This is also the site of the grouper/snapper fishery and the conch fishery, and it is increasingly a target for tourism and leisure capital development.

# Map of Puerto Rico: Municipalities and Key Sites

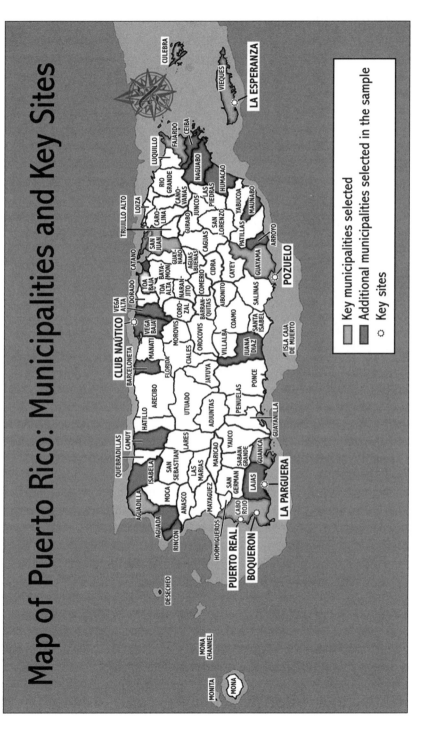

Legend:
- Key municipalities selected
- Additional municipalities selected in the sample
- ○ Key sites

2. *Aguadilla,* on the Northwest Coast, includes one of the island's most politically active fishing associations.
3. *Fajardo,* on the East Coast, is one of the most highly developed tourist areas on the island, home to several marinas, large resorts, small hotels, coastal reserves and a favorite beach, watersports, and general recreational location. As such, it is a site of potential conflict between commercial and recreational coastal industries.
4. *Guánica* is an area of incipient tourist development on the South Coast with little urban development and one of the more impoverished coastal populations.
5. *Vieques* is a twenty-mile-long island municipality off the East Coast of the main island, whose domination by the U.S. Navy, which owns all but the island's central corridor, has been a source of contention since the 1940s.
6. *Guayama* is a sugarcane-producing area on the South Coast, east of Ponce, similar to Mintz's (1956) Cañemelar. Though impoverished, it has been attracting new businesses, such as the petrochemical, plastics, and pharmaceutical industries.
7. *San Juan,* the seat of the island's government, is the most highly urbanized *municipio* and hence home to families of urban fishers.

In addition to interviewing fishers in these municipalities, we interviewed fishers in the fifteen randomly selected municipalities distributed over the island as shown in Table 1.1.

We set out to visit between fifteen and twenty fishing households in the seven targeted municipalities and between two and three in the randomly selected municipalities. In the latter, we

TABLE 1.1.  Additional Municipalities Included in the Study Sample

| North Coast | East Coast | South Coast | West Coast |
|---|---|---|---|
| Isabela | Ceiba | Arroyo | Aguada |
| Camuy | Naguabo | Juana Díaz | Rincón |
| Barceloneta | Humacao | Lajas | |
| Vega Baja | Maunabo | | |
| Vega Alta | | | |
| Cataño | | | |

attempted to sample randomly from the names that were supplied to us from the recently completed census of La Corporación para el Desarrollo y Administración de los Recursos Marinos, Lacustres y Fluviales de Puerto Rico (Corporation for the Development and Administration of the Lake, River, and Marine Resources of Puerto Rico), CODREMAR. Because of the vagaries of fieldwork and developments in fishing communities and associations that we considered of interest, combined with fishers' own fascinating life methods of moving among jobs and regions, our set of interviews and observations ended up more uneven than we had planned. In addition, although the bulk of the fieldwork was concentrated in the period from January 1988 to the summer of 1991, our work benefited from Valdés Pizzini's 1985 study of Puerto Real, two related studies of recreational fishing and coastal fishing infrastructure conducted from 1986 to 1987 (Griffith et al. 1988; Valdés Pizzini et al. 1988), and studies that focused on net fishers and the conch fishery conducted during the mid-1990s (Valdés Pizzini 1995; Valdés Pizzini et al. 1996).

Although we were primarily interested in fishers who moved between fishing and wagework, our discussion also includes information about Puerto Rican fishers who did not sell their labor. This information is important because full-time fishers' interactions with the marine environment, use of seafood markets, membership in fishers' associations and cooperatives, and conceptions of the sea and its gifts help establish the social and cultural parameters of those Puerto Ricans who move between fishing and wage labor. At the other extreme are individuals whose primary economic existence has been defined by wagework, who fish to supplement household subsistence needs, to earn additional cash, or to escape from the tedious routines and rigors of jobs in economic sectors such as manufacturing, agriculture, construction, and tourism. This variation raises the problem, similar to the difficulty in peasant studies, of trying to categorize fishers to reflect the full range of their behaviors (Kearney 1996). By examining the life histories of Puerto Rican fishers who are located at various points along this continuum—from full-time fisher to nearly full-time wageworker—we attempt to reveal how wagework and domestic production complement or frustrate one another, create or resolve tensions within fishing households, fragment communities, or help unite fishers in common causes and associations.

At the same time, we attempt to enrich Puerto Rican ethnohistory and ethnography by using the life histories of fishers as shadows, reflections, and critiques of historical processes and intellectual traditions that have influenced our thinking about Puerto Rico and Puerto Ricans. As workers, even workers who spend part of their lives at sea, Puerto Rican fishers are members of social classes and victims of labor market segmentation based on ethnicity, legal status, linguistic ability, gender, and other sociologically important criteria. Puerto Ricans—having been drawn by force into the orbit of North American influence at the end of the nineteenth century and granted U.S. citizenship in 1917 after prolonged racist debate in the U.S. Congress that ultimately allowed their conscription for military service during World War I—share with African Americans the reality of being minorities against their will. In this context, we pose questions about their political behaviors, whether subtle or overt, and their complacency about, acceptance of, subversion and manipulation of, and resistance against their positions in power structures and economic institutions. How self-conscious are they of their status? Do they agree with many of the ideologies that justify their use in servile, low-paying, high-injury jobs? Do they perceive clearly how their work in fisheries, like work among peasants, maintains and reproduces a reserve army of labor that helps keep wages low?

Yet as these workers move back to Puerto Rico, take up their nets and traps, oil their motors, and freshen the coats of paint on their boats, they move into a different theoretical light, becoming, as fishers, people whose cultural orientations are influenced by their time at sea. They interact with the natural environment in the political context of fishing associations and common property resources. Residues remain from their work on land. They remain workers, but now workers at sea, withdrawing from and thereby resisting the terms of capitalist production. Or, as many of us have argued at one time or another, they fall back on their own means of subsistence and thereby subsidize those for whom they work. Yet leaving wagework to fish, they run the risk of accommodating and having to resist alternative forms of domination, particularly those organized by merchant capital and the state. Resisting wagework, we illustrate in this book, can create divisions between family and community members similar to those that divide social classes from one another, even, paradoxically, while demanding coordination within households.

Thus, this movement between *la zafra* (the sugarcane harvest) and the sea, between factories and the sea, or between other hourly wage or piece-rate jobs and the sea is at once complementary and contradictory, generating ambivalent feelings and allegiances among those who undertake such treks and those who, as we do, structure intellectually, categories and theories about them. Social and cultural analysis is as likely to become entangled in the paradoxes of divided lives and selves as fishers who divide their lives between fishing and earning wages. Over the last years of the millennium, during the long, staccato process of writing this book, cultural anthropologists experimented with new paths toward knowledge as they maintained their central interest in local culture, local history, and social relationships within local, regional, and national processes. That we continue to be interested in local settings and processes—regardless of methods of observation, data collection, or presentation—lends some credence to the idea that local settings and processes define us in ways that the global corporation or the nation cannot, even in the face of cyberspace, transnational diasporas, or deterritorialized nation-states. Place is a social construction, and the social constructions of locality that we encounter among Puerto Rican fishers invoke the ways that they conceive of and represent coastal landscapes and offshore seascapes. Detailed attention to local settings among those who inhabit them leads to ethnic consciousness, questions of self and identity, and critical reflection on the various ways in which peoples from various cultural backgrounds and linguistic practices have been depicted by the media, politicians, and anthropologists.

It was from this enduring interest among anthropologists in local settings and local processes that during the last decade and a half of the twentieth century we entered the space of fishers and their family members and elicited information about their lifestyles. During these visits, we asked fishers and their families about specific features of their lives, especially events surrounding changes in their fishing operations and the deepening or diminishing nature of their experiences in their jobs. We also asked how they coped with crisis, how they enlisted the aid of family and friends or established ties with others in their communities. We were interested in their political views and behaviors, the ways they assembled their fishing crews and the dispositions of their catches. Throughout these interviews, we encouraged them to emphasize whatever they considered important, even if this meant straying from the structure imposed by our questions. Although we were interested in information on Puerto Rican fishers for its own intrinsic value, this work also attempts to weave Puerto Rican life histories into the biography of cultural anthropology.

## THEORETICAL BACKGROUND

The histories we present in this book invoke two fields of human experience that have formed much of recent world history: (1) domestic production, which consists of the practices of hunters, peasants, pastoralists, and others who hunt, fish, farm, raise livestock, or produce goods and services using family and household labor and (2) the incomplete incorporation of groups of domestic producers into capitalist systems of power, time, discipline, control, identity, and meaning. This incorporation has been uneven and highly variable. Puerto Rican individuals, households, networks, and communities have been differentially drawn into capitalist spheres of influence. They are not influenced merely as workers. Puerto Rican fishers, like domestic producers everywhere, also participate in capitalist economic systems as merchants, brokers, commodity producers, and entrepreneurs. Consequently, learning more about them forces us to revise common designations used in anthropology to talk about domestic producers, particularly the designation of "peasants" living in closed corporate communities (Wolf 1966), binary or hyphenated categories such as semiproletarian (Kearney 1996), or disguised wage laborers (Roseberry 1976). Puerto Rican fishers buy and sell fish, add value to their catch by processing their products, use credit systems, seek and obtain government licenses, and participate in political protests. They cannot be considered isolated to any degree or even fully independent in their production, since they neither inhabit villages in remote areas nor produce everything they need to live themselves. They inhabit, influence, and are influenced by the wider, complex societies of Puerto Rico, New York, Chicago, and the other locations, mostly scattered throughout the eastern United States, that have attracted Puerto Rican migrants and serve as anchors for the Puerto Rican diaspora.

Encountering groups that have been pushed and pulled by many influences has become common in anthropology and cannot help but become more common. By the end of World War II, many of the cultures studied by anthropologists could no longer be accurately represented without considering the ways the local histories of peoples and communities had become interwoven with world and regional history, a process that William Roseberry aptly calls "the internalization of the external" (1989:89). Especially important global processes included colonialism, independence and nationalist movements, European and communist bloc expansion, and the rise in the scope and influence of multinational corporations. More recently, anthropologists have turned their attention to ethnic conflicts, alternative histories, the growth of transnational communities, and other international currents of power and resistance (Anderson 1983; Basch, Glick-Schiller, and Szanton Blanc 1994; Roseberry 1989; Wolf 1982).

Yet, adept at and predisposed to studying small, local groups, few anthropologists have been willing to sacrifice attentive and detailed techniques of observation—the daily fare of fieldwork—for the broad, sweeping international analyses that too often rely on poorly collected statistical data. What would become of kinship, mating, ritual, foodways—of culture itself—under the influence of big projects? If at first our enterprise became one of balancing micro- and macroprocesses, or internal against external, later we became more concerned that we neither reduce local histories to the eddies and flows of international politics and economics nor venerate customs to such an extent that they are free of the influences of global change (Comaroff and Comaroff 1992, 1999; Mintz 1977). Working toward this balance has neither produced nor is likely to produce a standard method of data collection or analysis but has instead followed the path of any history: plodding, cluttered, dense with fashions and fads yet carried forward with occasional voices of insight, comparative work, and criticism.

We, the authors of this volume, came out of Marxist traditions yet entered a rapidly changing world of anthropology when we assumed university positions and began carrying on conversations in print. Through the first ten years of our careers, the issues of the 1970s elaborated by Immanuel Wallerstein and tempered by Sidney Mintz, Eric Wolf, Raymond Williams, William Roseberry, and others slowly withdrew into the shadows as the cultural studies and interpretive positions of postmodernism, having played themselves out in literary criticism, assumed importance in anthropology. However much we have experimented with these approaches, they have not satisfied our desire to illuminate the ways in which Puerto Rican fishers have been involved in a century of North American domination or our desire to describe and analyze (as clearly as possible and without either undue interpretive baggage or static measures of central tendency and dispersion) the local history, circumstances, and intrigues of fishing.

Prior to World War II, anthropological publications had noted that domestic producers sometimes exported labor to capitalist labor markets, but they rarely raised questions about how this might stimulate cultural change or alter local power or wealth relations. Thomas Gunn, writing about the Maya, stated:

> The Indians of British Honduras who live near settlements do light work for the rancheros and woodcutters; they have the reputation of being improvident and lazy, and of leaving their work as soon as they have acquired sufficient money for their immediate needs, and this is to some extent true, as the Indian always wants to invest his cash in something which will give an immediate return in pleasure and amusement. (1918:17)

Gunn's observations imply that, left to their own devices, indigenous peoples will never be reliable as wageworkers. What "ranchero" or "woodcutter," reading Gunn's account, would not think that the image of the "Indian" he profiles—the lazy savage, unwilling to work once he has had his bourbon and beans—calls for taking steps to make indigenous peoples more productive, more willing to work, or finding another labor force entirely? Gunn's assessment was by no means uncommon. In his classic "The Original Affluent Society," in *Stone Age Economics*, Marshall Sahlins points out that several observers of the leisurely habits of hunting and gathering groups have interpreted this as a natural propensity for laziness and indolence (1972:27–28). He quotes Martín Gusinde as follows:

> The Yamana are not capable of continuous, daily hard labor, much to the chagrin of European farmers and employers for whom they often work. Their work is more a matter of fits and starts, and in these occasional efforts they can develop considerable energy for a certain time. After that, however, they show a desire for an incalculably long rest period during which they lie about doing nothing, without showing great fatigue. . . . It is obvious that repeated irregularities of this kind make the European employer despair, but the Indian cannot help it. It is his natural disposition. (Sahlins 1972:28)

Observations such as these suggest that labor forces need to be constructed. They do not emerge as natural responses to job opportunities. Domestic production operations need to be either destabilized to the point of forcing some individuals into labor markets completely or eliminated, because as long as they exist, they deter people from putting in a full day of wagework. E. P. Thompson's (1974) work on the rise of the English working class, following Karl Marx, shows how the enclosure of common lands forced English peasants to seek work in manufacturing. Among fisheries, "managing the commons" of the open seas results in restricting access to natural resources in such a way that fishing families need to supplement fishing incomes with wage labor (McCay and Acheson 1987). Ann Stoler elaborates this point in her historical work on Sumatra. There Dutch planters debated the merits of different forms of labor control as though comparing blends of coffee, experimenting with several models and several ethnic groups before settling on East Indian coolies. Similar observations have been made by Philipe Bourgois (1989) and Mark Moberg (1992) in Central America, by Claude Meillasoux (1972) in Africa, by William Roseberry (1983) in Venezuela, and by David Griffith (1993) in the United States.

Various tensions exist between local groups—groups bound by descent, marriage, residence, production relations, common consumption habits, and other cultural factors—and the larger organizations of capital and the state. Often these tensions emerge because of the demands of domestic production: the home-based productive and reproductive tasks that help poor households around the world meet their subsistence needs. Because these tasks draw on family labor and, usually, investments of cash and hope, they can cause scheduling conflicts and divided loyalties. Employees may leave outside work undone to attend to home production needs, or home producers may neglect kitchen gardens or allow fishing nets to fall into disrepair because of the demands of their jobs.

Despite conflicts between domestic production and capitalist labor markets, well-organized economic and political regimes, historically, have encouraged domestic production in several ways. For example, to colonize new territory, governments may, through road construction and the provision of security systems, provide access to frontiers where domestic producers can establish small farms, fish, trap, hunt, or engage in other operations without fear of being victimized by criminals or populations perceived as hostile (Collins 1988; Stoler 1985). Merchants may encourage the growth of markets, using subtle and overt coercion, such as accepting codfish as currency or for payment of debts, to ensure the production of certain kinds of commodities (Roseberry 1983; Sider 1986). Or military regimes may expand their power to the point that domestic producers resist, fleeing into contemporary frontiers of night and illegal activities, such as smuggling, drug dealing, and cock fighting.

When domestic production practices conflict too sharply with capitalist labor needs, however, well-organized economic and political regimes have stepped in to restrict or curtail them. Methods of destabilizing domestic production are as diverse as those that encourage it, ranging from restricting access to the means of production to outlawing domestic production altogether. Small-scale producers are usually restricted from the means of production through privatization of resources and the slow or rapid concentration of ownership, facilitated by credit relations and various mechanisms of power. Fishers may be discouraged from fishing by laws that restrict the use of certain gear or access to the sea itself, through the creation of marine sanctuaries, privatization of the shoreline, or the tightening of licensing requirements (Griffith 1999; Valdés Pizzini 1990b). Domestic production may be outlawed entirely by declaring some of its aspects a hazard to public health, as occurred after the U.S. occupation of Puerto Rico, when home milk production, cigar manufacturing, and liquor distilling were outlawed (Picó 1986).

That domestic producers' practices have been discouraged in some places and times and encouraged in others has made it difficult for anthropologists to gauge their importance in terms of trajectories of capital accumulation, class formation, ethnogenesis, and other world historical processes. Under what conditions have domestic production practices, for example, stalled or sidetracked revolutions or frustrated class formation? Do they emerge differently during times of economic crisis, focusing wageworkers' attentions on consumption issues, than during times of economic growth? When can they be interpreted as forms of resistance? How do they reconfigure households and communities or influence conceptions of local history? How have national and international policies that regulate trade, labor, environment, foreign relations, food, and other factors influenced domestic producers' lives?

As if the difficulties that these questions pose were not enough, anthropologists studying in Puerto Rico face an additional challenge. The work directed by Julian Steward in Puerto Rico in the years after World War II occupies a central position within anthropological theory, method, and lore, similar to the Harvard project in Chiapas, Mexico; the Cornell project in Vicos, Peru; and the extensive series of bulletins produced by the Bureau of American Ethnology. In contrast to Vicos, Chiapas, or the bulletins, Steward's study broke from descriptive ethnographies of prewar, Boasian anthropology to establish an enduring tradition of drawing upon and revising theoretical propositions about processes of community formation and culture change.

*The People of Puerto Rico* (1956), a work coauthored by Steward, is to economic anthropology and cultural ecology what the Western Electric studies are to industrial sociology. This is in large part due to the influence of Steward himself. His early (1938) work among the Shoshone, Ute, and Paiute of the Great Plains traced relationships between technology and social organization, laying the foundation for an extended inquiry into the influence of ecological factors over demographic processes, especially group formation. The pioneering attempts of Steward and his associates to apply cultural ecology to the analysis of various community types in Puerto Rico demonstrated both the potential for an anthropological contribution to understanding complex societies and the weaknesses of a discipline nourished on studies of local groups (Steward 1955:chap. 12; Steward et al. 1956; compare Roseberry 1978). This work constituted early attempts by Steward and others to understand local change in terms of hemispheric domination that brought Puerto Rico into the United States in 1898 and the broader political economic developments that swept the globe after World War II. Steward's influential statement on cultural change ends by asserting that Puerto Rican "cultural nationalism" is "the spontaneous and inevitable reaction of all segments of the population to profound changes brought about by a set of institutions which has been imposed upon them from the outside" (1955:222; see also Duany 2000).

The words "from the outside" must have struck an ominous chord in the hearts of anthropologists who were used to thinking about fishing villages, peasant communities, bands of hunters and gatherers, and pastoralists as though they were isolated from the flows of world history. In ethnography, there had been a bias toward freezing people in time, discussing their cultures as though they had had no history, as though their traditions had remained unchanged for generations (Wolf 1982). We possessed the ethnocentric notion that Western, capitalist societies changed rapidly, whereas people who practiced production techniques that looked primitive to us had been stuck in some Stone Age. We embraced, all too uncritically, the notion that technological advances were the hallmarks of cultural change and that tradition lived in backward regions of wooden plows and animal-skin headgear (see Mintz 1985:24).

Steward's well-known attention to patterns of work organization and resource exploitation—the "culture core"—(Steward et al. 1956:6–7) reflected this bias, leading to a typology of Puerto Rican communities based on economic and ecological characteristics. His conception of the Puerto Rican countryside was one of farm towns where ways of life revolved around the rhythms of producing tropical crops. He deployed research associates in four agricultural communities on the island and in the upper-class circles of San Juan, encouraging Robert Manners's study of a tobacco and mixed-cropping community, Eric Wolf's study of a coffee-producing community, Elena Padilla's and Sidney Mintz's studies of sugar-growing communities, and Raymond Scheele's study of Puerto Rico's elite. The names of these research associates, as well as those of some of the field assistants on the project, are well known to several generations of anthropologists; they are associated with such enduring works as *Europe and the People Without History* (Wolf 1982), *Worker in the Cane* (Mintz 1960), and *Sweetness and Power* (Mintz 1985).

Steward's own contribution to the book was primarily typological, and in those chapters written by Steward and "The Staff," the authors disagreed over the utility of cultural ecology to explain the persistent poverty of large segments of each of the farm communities described at the heart of the volume (Roseberry 1978). Steward's analysis may have illustrated connections between family structure and the production of commodities, for example, but Steward was conceptually unprepared to examine the ways that production regimes subordinated families to their operations, encouraging some household structures while discouraging others. Instead, his notion of "levels of sociocultural integration" (Steward et al. 1956:14–15) distinguished between larger, smaller, and qualitatively different social arenas: family, band, tribe, and state. Yet this concept said little about the ways that those different social arenas influenced one another. Because his interests lay in formulating a general theory of cultural change, Steward's levels implied a developmental sequence, from primitive societies organized around families to complex states organized around institutions such as governments and corporations. Within this developmental sequence, if progress involved participating in ever more complex systems of organization, then those who failed to participate in wider social arenas must have been, in some way, backward, having fallen short of attaining the heights of complex societies that were capable, Vine Deloria reminds us, of setting their rivers on fire.

Steward was a fine anthropologist, and no good anthropologist would suggest that Puerto Ricans lacked the ability to achieve levels of integration on par with supposedly advanced nations of the world. Yet his model, similar to the traditional-modern and folk-urban models developed by an earlier generation of anthropologists and sociologists, implied as much. These models also implied that the most likely cause of technological inferiority was isolation—simply, peoples producing with primitive production technologies had not had enough exposure to outside influences to recognize the benefits of, say, outboard motors, freezers, the production regimes of the modern factory, or DDT.

Images of backwardness and isolation satisfied two demands in the social sciences of the postwar era. First, they legitimized the studies of anthropologists who continued to treat fishing communities, peasant villages, tribal groups, and others as though they were isolated, eventually spawning studies of the internal ecological and economic rationality of these groups. Second, they encouraged studies of modernization, spearheaded by sociologists and economists interested in diffusion of innovation. Studies of modernization inaccurately viewed increased wage labor among small-scale producers as symptoms of a general transition from preindustrial to industrial society, viewing social change as primarily a process in which domestic producers came to understand the benefits of advanced capitalism. These assumptions made it difficult for the researchers to examine ways that capitalist expansion could foster the reconstruction, maintenance, and reproduction of so-called traditional social, cultural, and technological adaptations (Roseberry 1989).

Anthropological analyses influenced by modernization theory (see, for example, Dalton 1971) focused on the internal logic of "traditional" or "folk" systems, paying little or no attention to the relationships that existed between local groups and national and international economic processes. These traditions guided much of the social science produced through the 1950s and 1960s, and a few anthropologists stubbornly continued to work from these perspectives into the 1980s and 1990s. Despite their failings, however, these approaches improved our understanding of the internal dynamics of cultural traditions and illuminated the constraints that cultural traditions placed on individuals as they attempted to adopt new technologies or new methods of organization. And they produced great quantities of data: community studies, ethnographies, regional analyses, case studies, and systematic overviews. Out of these traditions came such classics as Andrew Vayda's 1979 work on the Potlatch, Roy (Skip) Rappaport's enduring *Pigs for the Ancestors* (1969), Theodore Schultz's *Transforming Traditional Agriculture* (1964), and Edwin Spicer's (1954) and Benjamin Paul's (1955) edited volumes on the social and cultural consequences of technological change.

But as the high-water marks of these traditions produced lasting works, some who worked within these traditions justified widespread diffusion of technology along with forms of political domination that contributed to widening gaps between rich and poor (Hewitt de Alcantera 1976). By extension, they also facilitated political and economic developments, such as large-scale agriculture and aquaculture projects financed by government-backed loans, that created a demand for low-wage labor and partially destabilized fishing communities, peasant villages, and other centers of domestic production.

Ideologically, many who labored within the theoretical guidelines of modernization theory portrayed domestic producers as unprepared to deal with Western economic or political institutions, except possibly as workers or voters. The economic and political activities of domestic producers were represented as primitive or dysfunctional fetal gills waiting for astute scientists to fill them with breath and life. Such views, of course, were based on narrow and ethnocentric ideas of the exercise of power. Valdés Pizzini (1985, 1987, 1990b) has criticized this tradition while writing about the political activities of Puerto Rican fishers. In the process, he has documented a rich history of protest and effective lobbying, from disputes over the use of fish weirs early in the century, to confrontations with recreational fishers, to the organization of fishing associations in protest of coastal gentrification eighty years later. Instead of a linear progression, instead of the slow growth of political participation that some observers mistakenly view as evidence of modern politics (see, for example, Maiolo and Orbach 1981:7), Puerto Rican fishers' political behaviors have been irregular and episodic. The 1999–2000 colonization of the U.S. Navy lands on the island of Vieques, an effort spearheaded largely by fishing families, demonstrated that these political proclivities remain powerful, vibrant, and capable of garnering the attention of the international community.

While Puerto Rican fishers have demonstrated astute political skill at manipulating media and the support of established political leaders and parties, their bases of organization suffer from strong undertows, from within and without, that cause divisiveness within and among fishing communities. Fish dealers routinely compete for fishers' loyalties. Gear and territory conflicts pit families against one another. Labor market opportunities siphon off effective speakers. Crises of consumption within households initiate complex negotiations between genders and generations that dilute political participation. Such developments open fishers' eyes to the futility of much of the sustained political action that many consider the steady diet of effective modern politics. In light of the recent crises of nationalism, dissolving borders of so-called established nations, state-sponsored genocide, growing ethnic conflicts, and the absence of leadership and cooperation in formal political arenas, the opportunism and cynicism of Puerto Rican fishers' political activity seem mature. Sugarcane worker Don Taso's life history, translated and annotated by Mintz (1960), reveals similar opportunism and cynicism, making it difficult to sustain the argument that high levels of political activity are, indeed, modern. June Nash's recent comparisons among Mayan women, Bolivian tin miners, and unemployed General Electric workers suggests that, of the three, the U.S. workers had the most trouble objectifying their circumstances and attributing economic declines to shifting developments in the global political economy:

> Although General Electric workers have experienced in their own lives the movement of capital overseas and the destruction of the industrial base, I found a lower level of consciousness of their interdependence

in the global system than in Bolivia. Even in a recent interview of a few laid-off workers in 1991, I found that the majority justified the reduction in employment and layoffs of thousands in terms of the decline in profit. The repression of a critical political consciousness in the McCarthy era of community witch-hunts has borne its fruits in the narrow sphere of union organization and the pursuit of individual ends that dominate the workers' frame of reference. (Nash 1994:24)

Perhaps because of more sophisticated political understanding throughout the Caribbean and Latin America, the most trenchant criticisms of modernization theory and its offspring came from these regions (Beckford 1972; Frank 1967; Furtado 1976; Girvan 1973; Stavenhagen 1975; compare Kearney 1996). These criticisms called into question the typologies that justified the "system rationality" studies described previously. The central critics were from the structural dependency school, who first stressed analyzing dependent relationships between developed and underdeveloped countries and later championed world-system approaches (Frank 1967; Wallerstein 1974). Writing in the 1960s after touring William Faulkner's South and working for the Cuban press, Gabriel García Márquez—who was clearly familiar not only with Latin American scholarship and political thought but also with poverty and neglect in the supposedly developed world—summarized a core idea of this criticism in the following passage from *One Hundred Years of Solitude:*

> It happened once that someone at the table complained about the ruin into which the town had sunk when the banana company had abandoned it, and Aureliano contradicted him with maturity and with the vision of a grown person. His point of view, contrary to the general interpretation, was that Macondo had been a prosperous place and well on its way until it was disordered and corrupted and suppressed by the banana company. (1967:321)

García Márquez recovered the theme of North Americans misinterpreting Latin America in his Nobel Prize acceptance speech, which, quoting *One Hundred Years of Solitude*, questioned accounts that "historians had created and consecrated in the textbooks" (1967:322). Before the Nobel committee, in 1982, years after Latin scholars had made an indelible mark on North American anthropology, García Márquez proclaimed:

The country that could be formed of all the exiles and forced emigrants of Latin America would have a population larger than that of Norway. I dared to think that it is this outsized reality, and not just its literary expression, that has deserved the attention of the Swedish Academy of Letters. A reality not of paper, but one that lives within us and determines each instant of our countless daily deaths, and that nourishes a source of insatiable creativity, full of sorrow and beauty, of which this roving and nostalgic Colombian is but one more cipher, singled out by fortune. Poets and beggars, warriors and scoundrels, all creatures of that unbridled reality, we have to ask but little of imagination, for our crucial problem has been a lack of conventional means to render our lives believable. This, my friends, is the crux of our solitude. (1995: 134–35)

Between the time that Márquez wrote *One Hundred Years of Solitude* and the December of his acceptance speech, anthropologists and other social scientists had documented several ways that international political economic processes—markets, wars, the thirst for wage labor—constructed peasant regions and ethnic and tribal groups. In most cases, this fostered the surplus extraction of agricultural produce, fisheries products, rents, taxes, and labor. Celso Furtado's (1976) *Economic Development in Latin America* documented high levels of growth in Latin America during the Great Depression and World War II, when developed nations were too distracted to maintain old or create new dependent relationships with the nations of the third world. Rodolfo Stavenhagen's (1975) work on internal colonialism traced the ways that centers of power within Mexico undermined the opportunities for "acculturation" among peasants. Andre G. Frank's "Development of Underdevelopment" (1967) made similar observations. From the Caribbean came George Beckford's *Persistent Poverty* (1972) tracing the role of the plantation in controlling labor and organizing society, and Norman Girvan's (1973) work on the world control of mineral wealth by multinational corporations.

Anthropologists answered the challenges of these works by acknowledging their importance while criticizing them for their neglect of local history—or, in Mintz's words, "local initiative and local response" to the world system (1977). This was an exciting time to be entering the discipline of anthropology. The emphasis on local history created a pressing demand to listen to the voices of peasants, pastoralists, and gatherers—the voices, in short, of "others." Their views, behaviors, and actions criticize and reflect upon dominant social formations—states, corporations, political parties, unions, and so forth—even when these formations dominate them.

The social contexts of the 1960s and 1970s were instrumental in encouraging these developments. In 1965, England tightened its immigration policies and the United States opened its borders further, encouraging a shift of Caribbean migration from the British Isles to the United States: English-speaking peoples from the Caribbean began to move into labor markets and neighborhoods long the domains of Puerto Ricans, African Americans, and other citizen minorities. With improved civil rights, southern neighborhoods, especially in Florida, became more attractive to Caribbean immigrants. Developments such as the U.S. loss in Vietnam, Nixon's resignation, the dismantling of the British Empire, the deindustrialization of the United States, the large-scale poverty experienced in many U.S. cities and regions, homelessness, and despair encouraged anthropologists to question many of the principles upon which modernization theory rested. Throughout the world, anthropologists witnessed firsthand as development efforts failed in areas where they had had decades to succeed, further undermining modernization theory's credibility.

Studies produced during the late 1970s and early 1980s benefited from detailed accounts of the ways that capitalist relations of production organized peasant regions and communities through credit and marketing relations, simultaneously constructing labor markets that encouraged migration from underdeveloped to developed regions. These studies illustrated trajectories of global accumulation playing out in local settings (Brush 1978; Gudeman 1978; Orlove 1977; Smith 1977). At the same time, they cleared a path for theoretical interest in power and resistance (Chavez 1989; Nash 1994; Ong 1988; Scott 1985).

Studies of domestic producers moving into capitalist labor processes drew heavily on works about the changing international division of labor that accompanied deindustrialization in advanced capitalist nations and the growth of export platforms and fragmented production practices throughout the third world (Nash and Fernandez-Kelley 1983). This division of labor rests on economic segmentation and worldwide differences in workers' expectations and needs (Frobel, Heinrichs, and Kreye 1980; Gordon, Edwards, and Reich 1982; Sanderson 1985). Capitalist labor processes attract not only peasant fishers and other small-scale producers but also marginal workers such as students, youth, the elderly, labor migrants, minority ethnic groups, women, and unskilled or semiskilled "secondary" workers (Griffith 1993; Nash 1985; Stoler 1985; Gordon, Edwards, and Reich 1982). These workers often require lower wages to survive and reproduce because of cultural and social circumstances that have developed outside of or tangentially to the labor process. Peasant households, for example, provide portions of their own subsistence and reproductive requirements while they supply workers, on a long-term or seasonal basis, to wage-labor markets (Griffith 1983, 1984, 1985, 1986, 1987). Minorities and illegal alien labor migrants develop extensive networks to absorb the costs of seeking wage-work or moving where work is available, particularly to share housing and child care responsibilities (Fitchen 1992; Holmström 1984; Lomnitz 1977; Massey et al. 1987; Portes and Walton 1979). Women wageworkers often supplement their husbands' incomes with part-time, casual labor, taking jobs without benefits or opportunities for advancement (Benería and Sen 1982; Chaney and Schmink 1980; Deere 1983; Nash 1985).

Engaged in such practices, secondary workers often have lower expectations than primary workers regarding wages, labor unions, working conditions, and living standards. Because these employees are attractive to capitalist enterprises, capitalist businesses are encouraged to promote, through power and ideology, the maintenance of structural relations that perpetuate segmentation, marginality, and partial proletarianization. Despite these structural relations, anthropologists found that many people were skilled at circumventing the status quo and manipulating rules and refused to behave as disciplined workers or good little consumers. Largely in response to work that made it seem as though the expansion of capitalism reduced domestic producer economies to mere labor reserves, many of us found simple dominant/subordinate models inadequate (Cook 1982; Forman 1970; Griffith 1983; Orlove 1977; Portes and Walton 1979; Roseberry 1983).

Clearly, in some places and times, expanding capitalist relations of production stimulate change among direct producing
groups, occasionally resulting in the complete proletarianization
of small-scale producers (see, for example, Antler and Faris 1979;
Sider 1986). Yet uneven responses by domestic producers to capitalist expansion suggest that, at the very least, proletarianization
is occurring at a much less rapid and even pace than Karl Marx,
Friedrich Engels, and many economists predicted (Mintz 1977;
Long 1977). We have learned, instead, that alternative economic
arrangements or organizations—peasant fishers' and farmers'
households, networks, and neighborhoods, for example—have
become important theaters of both domination and empower-
ment. Within these arenas, people seek refuge and find sanctuary from domination, develop forms of resistance and revise
methods of political protest, and learn alternative ways of evaluating people, places, and things. In the process, at times, they
negotiate new methods of becoming dominated, having their sur-
pluses extracted, allowing their identities to be misrepresented
or changed.

Within these alternative lifestyles, too, people move into a different theoretical light. When they also happen to be fishers, we
begin to think about their interaction with the sea, their utilization of common property resources, and those subsistence and
market production practices that they share with peasants and
pastoralists. Like pastoralists, they range over territories that
belong to them only insofar as social relations, technical expertise, local history, cultural claims, and political circumstances
allow. Changes in one or more of these—difficulties of compos-
ing a crew, damaged gear, the annexation of waters by governments as bombing ranges or national parks—also initiate changes
in the extent of fishers' access to coastal and marine resources.

Like peasants, fishers draw upon their households and neighborhoods to compose their crews and organize for the common
good, value the fish they catch according to both its exchange
value and its use value as food, and humanize or decommoditize
many of their productive practices by linking them to the mental and social health of communities, the traditions of local history, and the long-term stewardship of marine resources.

Fishers live out critiques of conventional economics and sociological principles, question ideas that have long guided people who deal in social and economic policy, and present alternative models of economic behavior and social structure. Although we know that capitalist expansion can subordinate, restructure, or complement small-scale, household-organized production activities, we remain unclear about crucial dimensions of this process and how it reflects on the customary ways that economists, sociologists, and many other anthropologists interpret their data. Uneven participation in wage-labor markets by domestic producers suggests that local social and cultural factors mediate responses to the wider economic system. Local perceptions and understandings of social and natural environments influence which species or which pieces of technological knowledge enter capitalist circuits of value and which are routed into other spheres of production, subsistence, and exchange (Kopytoff 1989; McCay and Acheson 1987). Under what conditions do specific regions, communities, or households become marginalized, supplying primarily low-wage, menial, or seasonal workers to capitalist labor markets? How do other social units emerge as suppliers of specific types of low-wage workers, as the Kikuyu of Keyna (prior to Mau Mau) developed the reputation among the British as being particularly suited for civil service occupations (Stichter 1985; Worsley 1984)? Why are some households willing and able to use wage labor as a means of accumulating capital, whereas others are not? What are the impacts of these various responses to wage-labor processes on the real and perceived environments of small-scale producers? Finally, in our characterizations, should we think of fishers as resisting or subsidizing capital, or should we think of them in some entirely different way?

Puerto Rican fishers do not present simple answers to questions such as these. Fishing in the Caribbean has always been a more independent economic activity than peasant farming, in part because of the common property nature of the resource. And since 1898, Puerto Ricans have had access to job opportunities in the United States and the Caribbean, in addition to the plantation work that comes to mind when we think of the classic Caribbean rural proletariat. The story of Liche, for instance, gives a taste of the varied statuses that fishers hold and roles that they play on their life journeys:

> He could have had any name at all, but in his community bordering the sea, he was affectionately known as Liche. The personal history of Liche is that of many other men and women of the coast, whose evasive lives, detached from the dominant regimes of production, led them to the margins of the sugarcane fields. There they lived independent of—and inextricably tied to—the production cycles of sugarcane capital and all of those violent forces of *moira,* or destiny, that pushed them from the public housing projects of the coast to uncharted routes filled with Leastrygons and cyclopes.
>
> Like so many other coastal dwellers of this archipelago, when he was twelve years old, Liche learned, by silent observation and assiduous practice, the ancient craft of weaving straw fish traps and the more complicated art of weaving and knotting gill nets made up of three surfaces of mesh. At home, by the fish-cleaning shack, his mother repaired and wove these nets during the eternal nights and dawns as she awaited *papá*'s arrival, while Liche watched and learned to make his own fishing gear.
>
> Liche also learned to navigate the stormy waters of the Caribbean Sea and, like all the others, he learned the secrets of the bays, the reproductive cycle of the red hind grouper, the erratic movements of the dwarf herring. He learned to wield the encircling cast net and to perceive triangles everywhere as he looked out to the far-off hills, so that he might find his traps and lobster pots within a region most precisely described as every shade of blue. He learned to transform the mangrove forest, to make it his own, to use its wood for building houses and fish traps, to remove the dead boughs to make charcoal, to hunt water birds and gather their eggs, collect fresh oysters, and ambush the blue-gray land crabs.

Gradually infringing on the mangroves and coastal villages, salaried work in the fields came on with a seductive and binding force. The sugarcane industry absorbed the hands of those men and women in all phases of cane cultivation and sugar production. In *la bruja*, as all the Liches of the coast call the cane fields' fallow, idle period, only fishing offered solace. Fishing was therapy for the aching body, for the parts of themselves they had sold, therapy for the deep wellspring of dreams of independence and freedom from enslavement. That is when fishing became part of the legitimate equation of the region's political economy, sustaining a seasonal workforce and subsidizing sugarcane-related capital. That is how the world of sugarcane during the 1930s depended upon people like Liche, whose life history reflects entire communities living in the margins between the acquisition of capital and rural subsistence, flowing between the two, in a world of poverty and violence that pushed them further toward salaried work and modern living with every passing day.

Long before we did, Norman Jarvis interviewed fishers like Liche and struggled to understand the absence of full-time fishers in Puerto Rico, a recognized caste within the world of capital. In *The Fisheries of Porto Rico* (1932), he shows how fishers' productive lives were tied to labor as stevedores, cane workers, and other occupations, in complex employment strategies that continue to challenge social scientists. Jarvis asked himself the same question that Spanish officials had asked when they studied Puerto Rico's fisheries in 1803 and the same questions we asked, through the last decade and a half of the twentieth century, while we traveled to Puerto Rico's coast, talking with its fishers, digging up history.

Some call these fishers lazy individualists. Those who see them that way are the same people who want to control their lives, count them, explain them, define them for official purposes, and reduce them to their own categories (Scott 1985). But the lives and histories of these fishers defy our suppositions, as they trace a diversity of routes and pathways that we can barely grasp. Their lives, the lives of Liche and his fishing companions, were marked by a constant coming and going between salaried work, fishing, and activities that would classify them as "self-employed." They worked in the cane fields, in their vegetable gardens and on their farms, doing varied jobs, repairing machinery, building homes, and fishing. They worked as laborers in pharmaceutical firms and oil refineries; drove taxis in New York; waited on tables in hotels that are closed to people of their class; and sold fresh fish, fish turnovers, and fried or roasted johnny cakes made of wheat flour and coconut milk. Sometimes they worked just a few occupations. Other times they juggled a number at the same time, struggling to sustain their families and protect them from the fragility of their existence.

The masculine aura that surrounds fishing tends to hide (from male social scientists) the entire social universe that is controlled and governed by women in fishers' domestic circles. As Gudrun Haraldsdottir (1994) argues, the emphasis on the market and on the nexus of exchange as premier realms for social scientific analysis makes women invisible by theoretical assumption. Although industrial capital and factory work have worn away their participation, women traditionally controlled fishers' finances as well as the processing, preparation, and sale of the fish caught by their husbands, brothers, or sons. The registration and ownership of the fishing boats was also a woman's domain, as were net and trap repair. And if there was enough time, and if our proverbial Liche was absent, women abandoned the safer tasks to make a living at sea.

Liche, as we may have suspected, is a restless soul. His life story moves swiftly and at times takes a long circuitous path through space and time. Since the late 1940s and early 1950s, people like Liche joined the cadres of seasonal migrants who worked in the fruit harvests of the United States in a long-term repetitive cycle. In Puerto Rico, that pattern set a rhythm to the work and seasons that were rooted in their coastal homes. Some resisted the call to leave home. They stayed on the island, shifting, at times daily, between fishing and wagework: fire fighter and fisher, police officer and fisher, government employee and fisher—all within the framework of a perennial duality and logic that at times is difficult to comprehend. The most unsettling route of all was traced by the real, flesh-and-blood Liche of our tale:

After Liche learned to make his own gear, to divide his time between fishing and work on the land, one day, when he was still quite young, he left the island to struggle in other lands, to work, to raise a family, to send home the money he earned. Ten, twenty, thirty years passed: three decades during which he did not once fish or even see the ocean. But in a boat-building factory in New Jersey, in a bakery in the Bronx, on the production line of a chemical factory, on dozens of construction sites, every single day, he made a mental review of the waters, the isles, and the cays. The names of fish, their colors and shapes, formed part of a test to which he submitted himself daily. Amid the tedium of industrial production, in the imagination that factory workers retain, he would fly toward the island, think about and reconstruct the space and the people there, and entertain no thought unrelated to his return. Some actually break free one fine day and return, deproletarianized, to their home communities. Others, like Liche, are wounded and return to the island "disabled." They return to the sea, to fishing, to work in a job that offers therapy for their disability or for the alienation accumulated during so many years of mindless, repetitive labor.

The life histories presented here show fishers who, between times on the water, become farm workers in Connecticut, cooks in Miami, construction workers in Bayamón, and public works employees in Maunabo. Many have dwelt in Spanish Harlem in New York or in South Chicago. They have tasted the bitterness of life at the lowest watermarks of Western civilization and have experienced the beautiful compelling silences of the deep. They have loafed outside restaurant kitchens in Miami and sweated over jackhammers repairing highways in New Jersey. They have joined unions; struck; boycotted; and picketed; backed political bosses; protested working conditions; and helped earn more frequent breaks, fresher water in the fields, and higher wages. Yet they return, nearly inevitably, to fishing. They are not in the process of turning into a full-fledged working class, as Marx might have predicted or as modernization theorists might have desired.

If the conventional wisdoms lamented by García Márquez could not render the lives of Latin Americans believable in 1982, by 1990, when Octavio Paz, another Latin writer, accepted the Nobel Prize, he lamented the death of all those bodies of wisdom and thought that allowed, once, some fiction of rationality and common sense:

For the first time in history mankind lives in a sort of spiritual wilderness and not, as before, in the shadow of those religious and political systems that consoled us at the same time they oppressed us. Although all societies are historical, each one has lived under the guidance and inspiration of a set of metahistorical beliefs and ideas. Ours is the first age that is ready to live without a metahistorical doctrine; whether they are religious or philosophical, moral or aesthetic, our absolutes are not collective but private. It is a dangerous experience. It is also impossible to know whether the tensions and conflicts unleashed in this privatization of ideas, practices, and beliefs that belonged traditionally to the public domain will not end up by destroying the social fabric. Men could then become possessed once more by ancient religious fury or by fanatical nationalism. It would be terrible if the fall of the abstract idol of ideology were to foreshadow the resurrection of the buried passions of tribes, sects, and churches. The signs, unfortunately, are disturbing. (Paz 1995:263)

Moving between wage labor and fishing, between the dregs of capitalism and the beauty of the Caribbean Sea, who besides Puerto Rican fishers is better positioned to critique, with their ways of life, this transition from political and religious oppression and consolation to the impassioned and directionless lashing out of dislocated, alienated folk? Who is more prepared to negotiate the spiritual wilderness? At times we imagine the picturesque, romantic lives that people such as these fishers must lead, and we create our own visions of them, visions that may affect public policy and encourage its errors. When we glimpse the actual lives of these people, we are struck by their mobility, the way they freely cross the frontiers of life and work experiences. This very mobility may be key to our understanding of the fishers' combative style when they defend fishing grounds and coastal territories, for their fight is also a defense of their independence and their mobility as productive men and women. At the edge of an era marked by ethnic conflicts and fundamentalist claims to heart, mind, and soul, we can learn from these people—men and women who have suffered the welder's flame, the butcher's apron of blood; who have filled the open wounds of streets in our most treacherous cities; yet who returned to the villages of their mothers, the ways of life of their fathers, to take up their nets, their traps, their hooks, their lines, and fish.

# 2 Palatable Coercion

## Fishing in Puerto Rican History

THE OFFICIAL INCORPORATION of Puerto Rico and its people into United States hegemony was framed in the comments of Senator Vardaman made on the floor of the U.S. Senate on January 30, 1917:

> So far as I am personally concerned, I really think it is a misfortune for the United States to take that class of people into the body politic. They will never, no, not in a thousand years, understand the genius of our government or share our ideals of government; but the United States has taken this island; the investments that have been made there by American white men will induce the Government to continue to hold it. . . .
>
> I think we have enough of that element in the body politic already to menace the Nation with mongrelization; but if the Porto Ricans are going to be held against their will, as we are holding them now, then we ought to legislate for their interests. We should make the coercion as palatable as possible. (*Cong. Rec.* 1917:2250)

The condescending and reluctant tone of Vardaman's speech is no less prophetic than its paternalism. From 1898 to 1917, Puerto Rico had been treated as an occupied territory, ruled primarily by the U.S. Navy. In the words of another senator engaged in debate with Vardaman, Puerto Ricans were "in the most anomalous position that the people of almost any country were ever placed in; they are citizens of no country" (*Cong. Rec.* 1917:2253).

In the early twentieth century, what people could survive such a fate? Nationalism was becoming the prevailing mode of political organization, if one of the most elusive to social scientific analysis (Anderson 1983; Hobsbawm and Ranger 1983). During the first two-thirds of the century, people throughout the world learned to reckon their political identities as "citizens" of defined territorial and administrative units. Independence movements and revolutions established new states across the world, resulting in a fragile union of Soviet republics and allies while dismantling colonialism and establishing neocolonial relations between the United States and several newly independent nations. Paradoxically, as if the rush toward nationalism were little more than the air required to keep a flag unfurled, the last quarter of the twentieth century would be marked by people questioning national identities. Expressing allegiance in ethnic and other mythical terms, many waged wars to break away from nation-states into the "tribes, sects, and churches" that Paz lamented in his Nobel Prize acceptance speech (Basch, Glick-Schiller, and Szanton Blanc 1994; Paz 1995; Tambiah 1987).

What have these processes meant in the lives, homes, and gatherings of Puerto Ricans? History plays out inside everyday lives in curious ways, creating opportunities here and barriers there, suggesting hope to one person and despair to another. In an essay that appeared in an issue of the *Kenyon Review* dedicated to what we remember and what we forget, Susan Stewart describes a tenant shack on her family's property. It had been the locus of hope and failure for one of her cousins, who lived there with his young bride until she left him. Stewart writes:

> But in 1962, we knew none of this, not even pieces that might have fit into a story that itself might have been wound into the fabric of the past our parents and grandparents, aunts and uncles were always weaving. For we were a ruined family. We had been ruined by history, by fate, by bad judgment and bad weather—a family with a fall. And when you are part of a family with a fall you know who fought in the French and Indian War, and who fought in the wars after that, and who started a school, and who converted the Persians, and who could hear a piece of music once and commit it to memory—for these are the great dead of which you are merely a shadow. (1997:145–46)

This chapter attempts to locate history within fishers' households and lives, working through a few life histories with an eye toward how fishers might have utilized, more than simply experienced, historical trends that anthropologists and historians consider important. With this method, we cannot recover detailed accounts of what opportunities the U.S. acquisition of Puerto Rico in 1898 might have cut off or opened up any more than we can use the fishers we interviewed as voyeurs into the early haciendas. None of the fishers we spoke with could have remembered either period, let alone taken advantage of or suffered through the economic and political turmoil that inevitably occurred during those times. The oldest fishers we spoke with remembered only as far back as the 1920s.

Yet well-known and everyday moments stain social and cultural landscapes, and their residues linger and confront fishers even today. It is impossible, for instance, to describe any coastal adaptation in the Caribbean without discussing its relationship to sugarcane production, and fishing in Puerto Rico is no exception. Several of the fishers we interviewed worked in sugarcane fields during some part of their lives, just as Don Taso, probably the most famous Puerto Rican worker in the cane, at times in his life utilized resources from the sea. It is equally impossible to consider Puerto Rican history without considering migration to the U.S. mainland and the growth of Puerto Rican neighborhoods—what scholars today typically call the Puerto Rican diaspora—in major U.S. cities, primarily New York, Miami, and Chicago. The life histories we collected contain residues of these experiences, sometimes illustrating simply a link between sugarcane and fishing or migration and fishing, other times expressing a dynamic process of class formation and transformation, dominance and resistance. Because human memory is selective, portraying history through the lenses of life histories cannot be wholly linear or chronological. Thus, in contrast to typical renderings of Puerto Rican history, and in fact to most historical accounts, we move back and forth across time with the selective accounts of the fishers we interviewed. An explicitly chronological accounting would impose on the dynamic, nonlinear character of memory a structure that it does not always share with historical scholarship.

Among the first of these accounts comes from the life history of Alejandro Irizarry, who establishes links between sugarcane production, fishing, and migration to the mainland and who raises several questions of relevance to current theoretical issues in anthropology and interpretations of history.

Alejandro was born in Guayama, along Puerto Rico's South Coast, in 1918, and at age fourteen he began to fish for

shrimp in the rivers, a pursuit that many saltwater fishers differentiate from their own craft. Because in Puerto Rico
*pescador de agua dulce* (freshwater, or sweet water, fisher)
connotes laziness, when Alejandro married in 1934, at age
sixteen, he quickly learned saltwater fishing from his wife's
family. In 1938, he took seasonal work harvesting sugarcane,
working six months in fishing and six months in the cane
fields, a pattern that would last for sixteen years. In 1954, he
traded one seasonal farm labor job for another, leaving the
sugarcane harvest to take summer work on farms first in
Michigan, then in Pennsylvania, and then in Connecticut,
returning during the winter to fish with nets, lines, and traps.
Alejandro's lack of specialized fishing technique, though an
outgrowth of the general tendency toward gear and species
flexibility among artisanal fishers, also meshed well with his
migration schedule. It allowed him to enter and leave various
fisheries at various times of year, depending on the summer
agricultural schedules in Michigan, Pennsylvania, and Connecticut. Despite this appearance of synchronized scheduling,
which cultural materialists might view as adaptive, Alejandro
left this seasonal lifestyle during the mid-1960s and returned
permanently to Puerto Rico to take a public works job in
highway construction. This allowed him to develop skills as
a mason for the next six years, and, by the early 1970s, when
he was in his fifties, it allowed him to engage in fishing more
or less full time and still accept *chiripas* (part-time, casual
employment) in construction.

Alejandro's work history is an individual manifestation of
broader processes that first created a rural proletariat to work in
the sugarcane industry and subsequently drew portions of that
rural proletariat into the harvests of the mainland United States.
Yet it also hints at the relationship between the state and capital, particularly the state's role in freeing Alejandro from nearly
twenty years of seasonal work and cyclical migration when, in
the mid-1960s—at the height of the civil rights movement and
during the years when a disproportionately high number of Puerto
Ricans were serving in the Vietnam War—he returned home for

good. The Puerto Rican fishers' life histories illustrate time and again relationships among larger political developments and local peoples, relationships influenced by coastal ecology and changing seasons, along with ties of power, culture, and class. It is likely that combinations of politics and ecology influenced Caribbean peoples prior to recorded history as well. Prehistoric coastal adaptations in Puerto Rico and the Caribbean suggest that group fissioning and encounters with other aboriginal peoples, combined with pressures on inland sources of food, encouraged an early emphasis on the sea.

## PREHISTORIC AND EARLY
## COLONIAL COASTAL ADAPTATIONS

Natural shoreline formations and Caribbean marine resources have influenced Puerto Rican fishing since before the island was known by its current name. Taino peoples—Puerto Rico's aboriginal populations—relied heavily on protected estuaries, harbors, and river mouths, where they could establish fish weirs or set traps. Weirs function in much the same way as the fish pot or trap, the gear of choice among the majority of Puerto Rican fishers of the eastern and southern shores and one of the principal types of fishing gear used throughout the Caribbean. Weirs, like traps, trap fish, but since they are more permanently fixed than traps, they require a more definite claim of ownership over the resource.

Prehistoric occupation of the islands of Puerto Rico and the Caribbean depended on sea travel. The mariners who first settled the islands migrated up from the interiors of South America along the Orinoco River, crossing to Trinidad and Tobago and to Barbados and working north along the Lesser and then the Greater Antilles. Archaeologists have unearthed evidence that Arawak-speaking peoples moved up the Orinoco to the Caribbean coast a little over four thousand years ago. From this coast, Trinidad is visible, and once seagoing explorations from Trinidad discovered Grenada, the entire chain of the Lesser Antilles up to Puerto Rico was easily colonized, each island being visible from the one immediately to the south and east. The entire settlement process took around seventeen hundred years. The earliest Arawak pottery found in the Lesser Antilles is about twenty-four hundred years old.

The term "Arawak" has been widely misunderstood in the literature on Caribbean prehistory. This can be attributed, in part, to Christopher Columbus's confusion over where he had landed and the later tendency among Europeans to classify groups by hostility. Simply, Europeans considered anyone who expressed peaceful coexistence to be either Arawak or Taino, but they considered the more warlike peoples to be Caribs. Most of us know that Columbus died believing that he had found a western route to the East Indies, where people who were called Caribs had been encountered. This accounted, in part, for his confusion regarding the island peoples. Poorly transcribed versions of the word "Carib" in Columbus's logs, in which the "r" becomes an "n," form the etyomological root for the word "cannibal."

The aboriginal peoples of Puerto Rico are known as Tainos, the name derived from the word for "noble" or "good" in the language of the first Caribbean people encountered by the Europeans. We have no idea what these people called themselves, but linguistically they were Arawak speakers, most likely ethnically related to the Caribs. Little of their livelihood influenced present-day Puerto Rican culture, unless romantic images of Tainos as symbols of natural harmony or resistance to European domination can be counted.

The slow migration across the Antilles indicates the importance of the sea in the lives of Caribbean peoples. During the initial phases of settlement, between 2100 B.C. and A.D. 600, the Arawak seem to have settled in the interiors of the islands, establishing coastal villages only after they had put pressure on inland resources. The shift to coastal locations and increasing reliance on marine resources accompanied declines in populations of a land crab that provided a valuable source of protein. Early peoples relied heavily on conch, whelk, and several species of fish, gaining navigational skills while they fished. This migration was one of punctuated movement, characterized by spurts of exploration and frontier expansion followed by periods of more settled, horticultural ways of life, similar to the annual horticulture/hunting treks of contemporary South American tribal groups. Many

Puerto Rican fishers of today fish for long periods between intervals of seeking ways to make ends meet in low-wage jobs and returning to the calmer life of fishing whenever they can.

Although early peoples undoubtedly used fish weirs and other kinds of traps, along with spears and the gathering techniques of today's divers, George Foster (1969) claims that much of the gear in use in Latin America today came into use in the early years of Spanish occupation. Weirs, the principal gear to descend from the Tainos to the Spaniards, lay at the heart of disputes over fishery resources during the last decades of the Spanish domination of Puerto Rico and the first years following the Spanish-American War (Valdés Pizzini 1987). In this early period of U.S. domination of the island, Puerto Rican fishers showed their sensitivity to shifts in power and the opportunities they presented to common property disputes surrounding the use of weirs. Throughout this century, Puerto Ricans have drawn deeply on party politics and political posturing to achieve economic ends (Buitrago 1972; Mintz 1956). Fishers are no exception, marshaling their political skills toward struggles against sportfishing clubs, coastal gentrification, the creation of marine sanctuaries, and the U.S. Navy (see Chapters 7 and 8 in this volume).

Yet in the economy at large, fishing plays a modest, subservient role. The sea has always been most valuable to the Puerto Rican economy as a link to the rest of the world. Fishing is of small significance compared to defense, shipping, tourism, and other commercial and strategic uses of the surrounding waters; yet fishers early managed to develop an impressive inventory of gear. Along with weirs and fish traps, Puerto Rican fishers used various kinds of seines, gill nets, diving equipment and spears, and several hook-and-line rigs. It is difficult to say which of these types of gear was used most often or best defined Puerto Rican fishers. While a great deal of the writing about fishing peoples throughout the world defines fishers first by their region of origin and second by either the principal gear they use or the primary species they target, these defining principles do not apply to Puerto Rican fishers. Indeed, often these defining characteristics do not apply to fishers we know as Maine lobstermen and -women, Japanese trawlers, San Diego tuna fishers, Florida stone crabbers, Camurium canoe fishers,

North Atlantic purse seiners, or South Carolina shrimpers. Few fishers specialize too exclusively in gear or target species, especially fishers who work from boats shorter than forty-five feet.

Even where these names are accurate, the gear, regions, and primary target fish are only the most obvious and sometimes among the least interesting components of fishing peoples' craft. More generalized fishing strategies prevail in fisheries, such as Puerto Rico's, where the principal gear types capture a variety of shell-fish and demersal and pelagic fish and where the lack of abundance of any one major species throughout the year precludes specialization. Given these circumstances, it is misleading to identify Puerto Rican fishers primarily by the technical dimensions of their craft. Instead, a more accurate representation involves locating a few of the fundamental attributes of Puerto Rican fishing in its broader social context, particularly its link to the major force that has shaped Puerto Rico's coastal landscape over the past three to five hundred years: sugar.

## SUGAR AND FISHING IN PUERTO RICAN HISTORY

Sugarcane production in the Caribbean has been inextricably intertwined with fishing. On coastal haciendas and plantations, fishing played a key role in the subsistence strategies of the slave labor force. Using historical materials from Martinique, Richard Price (1966) documented that slaves with certain privileges were allowed to fish for their own subsistence, for the tables of plantation managers, and even for the slave markets that linked slave families across plantations and facilitated communication between major port cities and rural areas. Price contends that such privileges contributed to the development of adaptive skills and independence, which helped to ease the transition from slavery to freedom (1966:1364). Fishing offered a diversion from the labor control of the plantation. Fishing, perhaps heroically and mythically, constituted the axis of the formation of coastal communities, usually perceived as independent and vested with psychological and existential possibilities for improving one's standard of living. Similar developments occurred in other settings as well, where settlements that grew in coastal regions witnessed attempts by agricultural interests to restrict fishing because it offered workers some independence from wagework. Gisli Páls-son and E. Paul Durrenberger (1983) found this in Iceland, and David Vickers (1994) found it in Massachusetts.

Fishing initially helped slaves develop a parallel market system and trade information along with gifts and consumer goods, increasing contact with other individuals under the same labor conditions. This situation served the slaves well in their total rejection of the plantation system through escape, resistance, and the formation of maroon communities. For Price, fishing "served a function analogous to crafts and subsistence plots as a way out of the oppressive plantation system" (1966:1378). Puerto Rico was no different from other Caribbean locations. Coffee and sugarcane haciendas fiercely competed for labor against subsistence producer activities, small landholdings, and alternative economic pursuits such as fishing.

With the onset of manumission in 1873, coastal communities sprouted along the fringes of sugarcane plantations. In Puerto Rico, these represented subsistence enclaves populated by poor Creoles (people of mixed ancestry) and freed slaves from the early days of colonization. Fishing; charcoal production; swidden subsistence farming; animal husbandry; and a range of cottage industries and commercial activities, such as milk production and street vending, constituted the majority of production and exchange (Cardona Bonet 1985; Picó 1986). Where official and informal harbors developed, portions of the coastal population also participated in maritime occupations, all tied to the Spanish Crown's Seamen's Guild, a branch of the Spanish navy that regulated access to the sea. Only those who belonged to the guild were legally entitled to use boats to fish at sea.

Not until the end of the nineteenth century did the hacienda system begin to integrate wage laborers in great numbers. However, this transition was gradual, and proletarianization did not achieve a full head of steam until the U.S. occupation in 1898 and the advent of central sugarcane factories—or *centrales*—financed by U.S. capital (Giusti Cordero 1994). The agrarian economy of Puerto Rico during the first decades of the twentieth century was characterized by increasing proletarianization of the labor force;

the mechanization and modernization of sugarcane processing; the establishment of sugarcane central factories; and the concentration of the rural workers in company towns, agro-towns, and urban centers dependent on sugar. Sugarcane production at that time increased notably. From 1898 to 1930, production increased fourfold, reaching one million tons by the 1940s. In that decade, sugar was responsible for 40 percent of total employment, 20 percent of the gross national product, 20 percent of the amount paid in salaries and wages, and 62 percent of all exports (Heine and García-Passalacqua 1983:11).

Along with some material rewards, the Americanization of the Puerto Rican economy and society brought farmers and peasants the systematic dispossession of their land, subsequently turning them into rural workers (compare Steward et al. 1956; Valdés Pizzini In press). Forcing rural independent producers to migrate to cities or company towns, this stagnated rural employment growth—much of it because of declining coffee production—and increased labor discomfort; levels of pauperism; and, not surprisingly, disease, crime, and hunger. A series of rural and urban strikes shook the island from 1905 to 1940 (Heine and García-Passalacqua 1983; Picó 1983; Taller de Formación Política 1982).

Because of the poverty of primary sources, it is rather difficult to place into this historical perspective the social and economic trajectory of fishing households in Puerto Rico. Yet the available information suggests that fishers in the twentieth century came from two principal sources: The first group was historically connected with the Seamen's Guild, which represented maritime occupations and had important ties with the merchant marine and merchant capital in general, and the second group was from the peasant coastal settlements that sprang up along the fringes of harbors, coastal towns, and haciendas (Valdés Pizzini 1990b). From the life histories that we present throughout this volume, it seems safe to speculate that many of those coastal settlers were eventually incorporated into the ranks of the sugarcane workers. We also suspect that wage laborers became fishers to augment low-paying seasonal jobs that alternated between pronounced periods of high production (the harvest, or *la zafra*) and a waiting period (*el tiempo muerto*, or the dead season). During dead time (*la bruja*), many cane cutters remained idle, subsisting on advance payments and credit from the company store; yet clearly others pieced together a variety of jobs, hustled, scrounged, and performed whatever tasks they could to survive. These practices, developed in the interests of large-scale agricultural production, included fishing and were not always viewed in the most positive light by those in power. In his *Report on the Civil Affairs of Porto Rico*, General George W. Davis, the first naval commander of Puerto Rico after the U.S. occupation, wrote:

> The class who can't fulfill [the literacy requirements to vote] . . . are usually in a state of abject poverty and ignorance, and are assumed to include one-fifth of the inhabitants. They are of the class usually called *peones*. This word in Spanish America, under old laws, defined a person who owed service to his creditor until the debt was paid. While those laws are obsolete, the conditions of these poor people remain much as before. So great is their poverty that they are always in debt to the proprietors or merchants. They live in huts made of sticks and poles covered with thatches of palm leaves. A family of a dozen may be huddled together in one room, often with only a dirt floor. They have little food worthy of the name and only the most scanty clothing, while the children of less than 7 or 8 years of age are often entirely naked. A few may own a machete or a hoe, but more have no worldly possessions whatsoever. Their food is fruit, and if they are wage earners, a little rice and codfish in addition. They are without ambition and see no incentive to labor beyond the least that will provide the barest sustenance. All over the island they can be seen to-day sitting beside their ruined huts, thinking naught of tomorrow, making no effort to repair or restore their cabins nor to replant for future food. The remarks of Mr. James Anthody Froude in his work on The English in the West Indies apply with full force to these people:
>
> > *Morals in the technical sense they have none, but they can not be said to sin because they have no knowledge of law, and therefore they can commit no breach of the law. They are naked and not ashamed. They are married but not parsoned. The women prefer the looser tie that they may be able to lose the man if he treats her unkindly, yet they are not licentious. . . . Their system is strange, but it answers. . . . There is evil, but there is not the demoralizing effect of evil. They sin, but they sin only as animals, without shame, because there is no sense of doing wrong. They eat forbidden fruit, but it brings with it no knowledge of the difference between good and evil. . . . They are innocently happy in the uncon-*

*siousness of the obligations of morality. They eat, drink, sleep, and smoke and do the least in the way of work they can. They have no ideas of duty, and therefore are not made uneasy by neglecting it.*

Between the negro and the *peon* there is no visible difference. It is hard to believe that the pale, sallow, and often emaciated beings are the descendants of the conquistadores who carried the flag of Spain to nearly all of Spanish America, and to one-third of North America. One family of industrious people, such as are found all over the United States, contributes more to the general prosperity and wealth of the country than ten families of these peones. (1899:18)

Whereas analyses of the Americanization of the sugarcane industry in Puerto Rico have emphasized the massive proletarianization of the peasantry and the incorporation of laborers into full-fledged capitalism and the world economy, our work suggests that the sugar *centrales* actually furthered the semiproletarianization of Puerto Rican labor. This uneven proletarianization allowed the *centrales* to remain profitable, paying rather low annual wages because the labor force assumed a good portion of its reproductive costs as members of cane workers' households engaged in various activities to support their families. Many scholars have argued that peasant communities serve similar purposes around the world (Griffith 1985; Scott 1976; Wolf 1966).

Fishing was central to productive activities. In her essay "Nocorá" in *The People of Puerto Rico*, Padilla described an array of subsidiary activities that sugarcane workers relied upon, including casual work or varied jobs, domestic activities for other households (laundering, seaming, cooking, cleaning), moonshine rum production, gardening, sand hauling, gambling, and fishing. Migration also emerges as an important strategy:

Others migrate to San Juan or some other city in the island to find a job. In the past few years, specially since the end of World War II, many young men have migrated to the United States a number of times as contract labor to work in the agricultural harvest.... Most migratory agricultural workers, however, return to the community during winter when the sugar harvest begins. Men with families cannot travel to find work, and they stay in rural Nocorá. (1956:285–86)

This pattern remains today, except that the slow demise of sugarcane production forced even men with families to migrate into the same labor markets indicated by Padilla: New Jersey, Washington, New York, and Michigan.

Mintz also discusses the role of subsidiary activities, often through the laborers' own words ("One must live illegally here" [quoted in 1956:361]), and he joins Padilla in pointing out the importance of fishing. However, Mintz makes an important distinction. In Cañemelar, the community where Mintz worked, both agricultural workers and full-time specialists fished. The passage in *The People of Puerto Rico* deserves full quotation, as it illustrates the pervasiveness of the *centrales* in subsidiary activities—that is, in the semiproletarianization of Puerto Rican labor:

> Fishing is a year round activity which supplies full-time employment for about six adults in barrio Poyal. Yet even these full-time fishers will be found sometimes working in the cane during the height of the harvest. Fishers range from the full-time operators of sailboats . . . to those who fish from rowboats or from the shore for sport or for food for the family. Sailboats represent a very substantial investment of cash—about two hundred dollars. Because they enable the fishers to use their nets and traps several miles offshore, and because of their large carrying capacity for the catch, sailboats are an enviable possession. But the cost of a sailboat is prohibitive and no sugar cane worker can hope to save enough to buy one. Besides, operating and maintaining a sailboat involves special skills not known to most local people. Full-time fishers have a special status in the barrio. Their cash income is perhaps less than that of an energetic *palero* or foreman, but their boats represent a significant accumulation of capital and their skill and knowledge are much admired. When these fishers are not at sea, they spend their time mending and making nets and traps, caulking their boats, repairing sails, and otherwise renovating equipment. A much more important group of fishers numerically are those sugar cane workers who fall back on their fishing skill during the slack season in the cane. These men, about forty of them, fish from rowboats or along the shores and use their catches for food, to maintain social relationships via gifts of fresh fish, or to provide a small extra cash income. (Mintz 1956:362)

Padilla and Mintz document that, during the hegemony of the sugarcane *centrales*, fishing was a subsidiary activity for rural laborers to earn additional income or fill their tables with high-protein foods. Although it helped coastal households reproduce themselves while providing labor to capital, the growth of a fishing sector was not entirely beneficial to capital. From capital's perspective, fishing subsidized workers' households as it potentially created problems of labor reliability, since some workers were able to gain enough expertise to fish full time, creating an alternative to working in the cane. In North Carolina, Griffith (1999) found that African American women who provided labor to the blue crab industry, where work was seasonal and was irregular even during the season, not only developed alternative income-generating activities but also responded to new economic opportunities in their communities and eventually failed to reproduce the labor supply to crab processing. Along with other economic activities, some of which fall in that shadowy zone of the informal or illegal economy, fishing contributed to the subsistence of proletarian households, while sugarcane wagework slowly initiated the households of independent producers (such as fishers) into semiproletarianization. Distinctions between the two types of workers—fishers and cane workers—rest on the time spent in each production regime, along with the capital accumulated and invested in boats and gear, which was, and still is, a sizable amount. In the 1940s and 1950s, the cost of a sailboat represented more than the average yearly earnings for a rural worker (Valdés Pizzini 1985). Although lack of skills, capital, and time precluded many people from entering fishing, many turned to independent production of marine commodities rather than resort to work in the sugarcane *centrales*.

Since 1930, fishing, like other domestic production activities, battled with sugarcane production for labor, space, and household commitment. Jarvis, a U.S. fisheries agent who studied the fish markets of Puerto Rico and the U.S. Virgin Islands, found that of 1,403 men who engaged in fishing, only 600 could be classified as full-time fishers:

> The great majority of fishers in Puerto Rico depend on plantation work, employment in the sugar centrales, or stevedoring at the docks and landings as much [as] or more than they do fishing. Fishing is followed as a sole occupation only where their work cannot be obtained or the demand for fish is fairly extensive. (1932:14)

Those areas where Jarvis indicated full-time fishing coincide with the major regions sampled in our life history study: areas with histories as official and unofficial harbors and trading centers on the island. Again, we point to a historical tie between maritime occupations; merchant capital; and the reliance of the labor of those areas on, among other activities, part-time or full-time fishing. One of Jarvis's key findings was that the number of fishers fluctuated by season. From June to January, dead time in sugar, the number appeared to increase. Jarvis suspected that in some areas, such as in San Juan, few men who declared they were "full-time" fishers fished only "at intervals between loading ships, or to supplement other irregular employment" (1932:14).

Jarvis explained the dichotomy between full-time and part-time fishing as a matter of the individual as guided by market conditions and economic rationality. Jarvis argued that the fish market conditions were detrimental to the fishers, since fish often spoiled, and "no regular channels exist for the distribution of the catch" (1932:14). Although that argument may have been partially true, it appeared that policy decisions were to blame in part as well. Jarvis noted that "the condition of the fishers has not improved as was expected under the American administration" (1932:15). Referring to older informants, and comparing notes with the 1899 Wilcox report on the island's fisheries, Jarvis recognized that, although retail fish prices increased, ex-vessel prices did not rise in the same proportion, leaving the fishers with "a small margin for the replacement of boat and gear" (1932:15). Yet proletarianization, according to Jarvis, had some benefits for the transformed fishers: "Working ashore, he [the fisher] is paid regularly, while the hours are shorter and the work often easier" (1932:14). This may have been the case in some jobs, but not in the job most commonly available: working in the cane.

What must be stressed here is that fishing, along with other independent-production activities, competed with sugarcane production for the labor and sympathies of household members along the coast: As opposed to having a symbiotic function, sugar production and fishing incorporated households into coastal political economy differently—the former proletarianizing coastal dwellers, the latter negating that process. Public policy encouraged these activities unevenly, promoting fishing during times of severe economic dislocation while promoting sugar production more or less all the time. Similarly, in Iceland, deproletarianization involved households that had been partially or completely incorporated into wage-labor positions yet used fishing to leave these jobs (Durrenberger and Pálsson 1989). Caribbean ethnology has documented a vast interest among sugarcane laborers in escaping plantation work and dodging proletarianization, in some cases with fishing being their principal means of escape. Specifically, following Price's contention that fishing served as an alternative to the plantation system, anthropologists have discovered that fishing and occupational multiplicity conform to a labor strategy of survival and semiproletarianization in a world dominated by capital and poorly paid wagework (Comitas 1974). Caribbean people divide labor into working for oneself (as in fishing) and working for others (as on plantations, in government, or in other jobs). Fishing as an independent-producer activity invokes creative skills, in sharp contrast with plantation labor, which is a primary reason that fishers in Puerto Rico speak of fishing as therapy (see Chapter 5, this volume; see also Gatewood and McCay 1990; Mintz 1956). On the side of production, fishing also allows landless or small-scale landowners to support their households without having to engage in full-time wage labor, as has been documented throughout the Caribbean (Valdés Pizzini 1990a).

It is clear, as well, that producers in Puerto Rico prefer fishing to working on plantations. Sugarcane plantations, vis-à-vis fishing, represent slavery, foremen, and wage-labor time and task discipline, while fishing represents independence and control over oneself and the production process and is capable of providing a satisfaction not found in wage labor (Gatewood and McCay 1990; Levy 1976; Valdés Pizzini 1985, 1990a). This view correlates with surveys that find that fishers in Puerto Rico fish, among other reasons, because they enjoy the contact with nature and the capability of structuring their own time and efforts that it provides (Gutiérrez Sánchez 1982). Other studies have shown that fishing among Puerto Ricans produces "spiritual and psychological" benefits, including pleasure, freedom, a flexible time schedule, and "no bosses" (Blay 1972:64–65). Similar responses have been elicited from fishers in other areas (Gatewood and McCay 1990; Griffith 1999; Pollnac and Poggie 1978). The psychological benefits of fishing may have cushioned the impact of the eventual decline of sugar from its central place in the island economy—both for the fishers themselves and for the state, as it encouraged fishery development to offset the hardships of economic dislocation.

## THE DEMISE OF SUGARCANE PRODUCTION

After 1940, the Commonwealth of Puerto Rico responded to the consistent weakening of its sugarcane industry with a development and modernization plan. U.S. capital poured in as transfer payments and industrial development. In 1950, sugar production increased, but the increment in wages and the drop in demand for sugar doomed the future of the former king of the island's economy. During this period, South Florida sugar producers, competing with Puerto Rico, benefited from both the British West Indies Temporary Alien Labor Program and from technology developed by the U.S. Army Corps of Engineers. The labor program allowed Florida producers to import exceptionally docile labor from Jamaica, St. Vincent, St. Lucia, Dominica, and Barbados, while the Army Corps of Engineers built one of the most complex water control systems ever known. Puerto Rico faced additional competition from Cuban producers and from the development of other sanctions, such as beet and corn sugars (Mintz 1985).

The commonwealth addressed the crisis by buying the collapsing *centrales* from local and U.S. companies and *hacendados*, forming the Commonwealth Sugar Corporation to assume the bankruptcy and demise of the industry. The commonwealth's development strategies also stimulated the demise of the agricultural export economy, creating industrial and commercial jobs in the urban centers that attracted large numbers of peasants and rural proletarians to the cities. Despite this growth, the economy was and still is unable to cope with the demand for jobs in all sectors of the economy. The accretion of fertility rates and population growth triggered by improved medical services and dietary patterns augmented large numbers of unemployed rural proletarians and lumpenproletariat who flocked into the cities and migrated to the U.S. mainland (History Task Force 1979:141; Kearney 1996; Picó 1983:chap. 4).

Declining economies of the northeast cities of the U.S. mainland, fiscal crises, housing problems, and crises of political legitimacy joined with an expansion of the welfare state, fertility control programs, and the massive transfer of federal funds and aid
to the poor of Puerto Rico (Bonilla and Campos 1985; Pratts 1987).
The transfer of assistance fueled the process of return migration
to the island. Many fishers, including Jason Cruz, participated in
this trend:

> A fisher from the southern coast, an area affected by high
> unemployment, Jason worked loading sugarcane in the wag-
> ons in the Central Azucarera from 1948 to 1955. In 1956, he
> began a long history of jobs that included working in New
> York on cauliflower farms, migrating consecutively until
> 1959. Returning to Puerto Rico that year, Jason worked in
> construction, helping to build a petrochemical plant. The
> construction industry was at its peak in the 1960s, mainly
> because of an increase in housing projects and the construc-
> tion of industrial infrastructure. Using affinal relatives and
> cousins, Jason was able to access jobs in that sector. With a
> *compadre,* Jason worked as a mason on housing projects
> along the South Coast. After the construction boom ended,
> Jason traveled to Massachusetts to work for six months on
> tobacco farms, employment found through the local office of
> the Department of Labor. During the months that he spent
> back in Puerto Rico, Jason worked in an auto parts plant and
> then in a hospital over a period of nine months and then
> returned to New Jersey to work in a light fixture factory.
>     The labyrinth of proletarianization in agriculture, the auto
> industry, farm labor in the United States, and construction
> finally came to an end in 1976, when Jason was able to begin
> to fish full time—this time for therapeutic reasons.

Throughout Jason's life, the local government depended, and has
continued to depend, on federal funds and programs to alleviate
economic crises on the island. The decline of Puerto Rican coffee,
tobacco, and subsistence farming; land accumulation on behalf of
the U.S. companies; government land expropriation by the armed
forces; and social displacement of coastal settlements all altered

the life histories of households along the coast and in the highlands. They left in their wake a social soup simmering with sanctions, state suppression, violence, strikes, and labor disputes.

During the late 1930, 1940s, and 1950s, the state suppressed incidents of social, labor, and political unrest through a mixture of armed responses and various social programs. Programs included the Puerto Rico Emergency Relief Administration (PRERA) and the Puerto Rico Reconstruction Administration (PRRA), which provided between 1933 and 1941 more than 230 million dollars in social and economic programs for housing, food, and clothing for the rural poor. Still today, the acronyms PRERA and PRRA evoke recollections of harsh times, state benevolence, and the demeaning household state of abject poverty and helplessness. Industrial development was indeed structured, among other reasons, to draw rural workers gradually into the hearth of industrial production, thus minimizing the dislocating effects of the demise of export agriculture and sugarcane decline. But the ultimate solutions to the socioeconomic and political problems in the fields and urban centers were migration, female sterilization, and other family planning strategies (Bonilla and Campos 1985). The commonwealth officially described migration as the escape valve for Puerto Rico's problems and made subtle but effective efforts to keep this valve open, sending migrant farmworkers to the mainland and to semipermanent settlements in the industrialized cities of the Northeast Coast of the United States. Agricultural producers all along the Eastern Seaboard—but particularly in New Jersey, New York, and New England—happily hired Puerto Ricans, whose reputations as hardworking *jíbaros* preceded them (Griffith et al. 1995).

Inevitably, Puerto Rico's fisheries, in the twentieth century, turned into a labor buffer zone, in which members of the displaced, seasonally unemployed proletarian population found food, recreation, and cash to maintain their families during the dead time, which they call *la bruja* (the witch). It is thought that, during the 1940s and 1950s, the number of fishers increased notably, because of unemployment across the island. Most of the fishers we interviewed grew accustomed to the seascape in coastal settlements along the fringes of the sugarcane *centrales*. In this developing culture of *chiriperos* (people who take many varied jobs throughout their lives), learning to use marine resources for the benefit of their households was an essential part of their life experience. The mangroves, lagoons, reefs, and platform habitats provided them with a training ground for a trade that could help them solve economic problems. This is a rich and diverse seascape and topography, seldom fully understood by nonfishers. But it is fully ingrained in fishers' memories and cognitive schemes, constructed in a communal learning process, and refined over the years despite time and distance.

Mintz (1956) observed in Cañamelar that becoming a full-time fisher presented insurmountable hurdles for rural workers, as the cost of sailboats, which represented significant capital commitment, was prohibitive for workers in the cane. Because time committed to sugarcane production also precluded wage laborers from engaging in full-time fishing, fishing became a subsidiary and seasonal activity, pursued most vigorously during dead time. Low prices for fish, lack of refrigeration facilities, and minimal capital investment made it difficult to pursue fishing as a full-time livelihood. Government officials who evaluated the fisheries during the first part of the twentieth century agreed that, in light of ex-vessel prices and marketing structure, fishers' earnings were quite meager and did not allow for an adequate subsistence or intensive capital investment (Jarvis 1932; Vélez, Díaz Pacheco, and Vázquez Calcerrada 1945).

As sugar continued to decline following World War II, in response to world prices and the emphasis on industrial development and changes in patterns of land use, the prices of fish and shellfish showed sustained increases during this period. Coincidentally, government programs for the reincorporation of labor into other sectors of the economy were initiated, supporting sugarcane workers and promoting efforts to develop the local fisheries. It may have been the decline in sugar and the growing state interest in investing in fisheries that underwrote, in part, the growth in the number of fishers from 1930 to 1950, followed by two decades of decline (see Figure 2.1).

In line with the strategy for development, the commonwealth initiated efforts to increase the production and improvement of

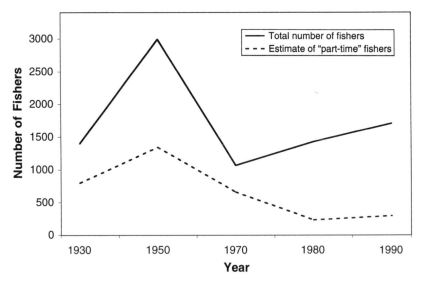

FIGURE 2.1.   Full-time and part-time fishers for Puerto Rico as a whole.

marketing facilities and fishers' opportunities at the moment in which prices increased (Iñigo 1968). Government efforts included the motorization of the fleet through loans for the purchase of boats, the simultaneous creation of the fishing credit section at the Department of Agriculture, the construction of centers for the sale of gear and equipment at low prices, the provision of fishers training courses, and the construction of landing centers called *villas pesqueras*. Many of the fishers we interviewed started buying motors and boats only when capital was available through government development programs. The development of refrigerated facilities and motorized fleets promoted uneven growth in the industry, including the large Puerto Real fishing fleet (discussed at the end of Chapter 3).

This rather uneven, discontinuous, and haphazard development effort reveals the loyalty of the Puerto Rican government to other productive forms and processes, such as manufacturing. It is rather telling that the commonwealth invested considerable time, effort, and funding to prime distant fisheries (such as the tuna fishery, based mostly in the Pacific) in order to provide raw material for U.S.-owned tuna canneries and the Newfoundland codfish industry, which supplied the main staple—saltfish—for Puerto Rican laborers (Giusti Cordero 1994; Valdés Pizzini et al. 1996). These historical decisions, and the relative paucity of state investment in local, artisanal fishing, led Ricardo Pérez (2000) to argue that state intervention actually retarded fishery development, keeping local fisheries small-scale, artisanal, household labor–based, and inextricably tied to the workings of capitalist enterprises that recruited labor for their operations. As this uneven growth suggests, success in fishing and capital accumulation at the household level (as subsequent chapters show) depended on the array of activities in which relatives, the state, and community members became engaged. Such social processes occurred against the ethnographic background described in the chapter that follows.

# 3    Puerto Rican Fisheries

A CASUAL DRIVE along Puerto Rico's coast reveals the diverse fruits of the island's fisheries. Nearly every *villa pesquera*, or fishing association, sells the seafood pastries known as *empanadillas de pulpo, de langosta, de camarones, de chapín*—the Caribbean pies made with octopus, lobster, shrimp, or little box-shaped trunkfish. In Guayama, on Puerto Rico's South Coast, vendors selling from a van on a side road between two pharmaceutical factories offer fresh mackerel steaks marinated and fried in sweet onions and olive oil, keeping them warm in a glass case fitted with five sixty-watt bulbs. Across from their mobile restaurant stretches a sugarcane field, and beside the van they set up tables and chairs beneath a cluster of flamboyant trees. Factory workers drift up to the side of the vehicle to buy midday meals of the seasoned fish served with rice, beans, and deep-fried slices of plantain. The woman and man selling mackerel may have spent the previous weekend at Las Croabas, in Fajardo, on the island's East Coast, competing with women who set up even more temporary roadside stands to sell octopus and conch salad in plastic cups. To create this delicacy, they dice the octopus tentacles or cube the conch and season it with halved green olives and the same sweet onions and oil that season the mackerel.

Across the island from Fajardo, beside a bay in the northwest municipality called Aguadilla, fishers hang tuna, bonito, and dolphin from crossbars erected in the shade of a makeshift pavilion of palm fronds, selling scaled, cut fish in plastic bags. Farther south, taking advantage of the weekend traffic, a man sells *pinchos de tiburón*—shark shish kebabs—alongside the road between Mayagüez (the major metropolis on the West Coast) and Puerto Real (home to the island's largest fishing fleet). Both cities face the wide, treacherous Mona Passage, one of Puerto Rico's deepest and most productive bodies of water.

Yet exploitation of the island's marine resources does not always depend on large, deep, fertile bodies of water. East of Aguadilla, for example, near the coastal urban center of Arecibo, one can purchase tamales made from yucca and *setí*, a marine larvae of the freshwater fish known as *chupapiedra*, or stone sucker (*Sicydium plumieri*). The larvae's oversized eyes peek out of the mashed yucca like the eyes of fish pulled from a deep-sea trench. On weekends, a teenaged boy heats these tamales on a thin sheet of metal over a low fire and sells them to tourists. During the week, he and his father rise early to tow mosquito netting in the river mouth for the *setí* as they migrate into the river. They take them home for the boy's mother to mix with yucca and form inside banana leaves. From this commercial pursuit, the family can earn up to two hundred dollars per week.

This cottage commerce, these sweetmeats are only the most obvious products of Puerto Rico's fisheries. Although they testify to the multiple ways that Puerto Ricans utilize the resources of the Caribbean Sea, they mask other products of the fisheries. Less obvious than these fruits enjoyed in seafood restaurants by residents and tourists are the years of effort of the sons and daughters of fishing households, their knowledge of the Caribbean ecosystem, the craft of the seafarer, gear construction, the routine maintenance required to harvest fish and shellfish, and the expressions of love for the water that one can glean from talking with Puerto Rican fishers.

The fisheries are distinguished by how much variety emerges from such a seemingly simple pursuit. All along Puerto Rico's coast, wherever urban sprawl has spared a short stretch of beach, a sheltered inlet, or a bay fitted with pilings and piers, small wooden fishing craft lie upside down or ready to be launched into any one of dozens of marine activities. These vessels are usually short, almost stubby looking; blunt at the stern; and worn where someone, daily or weekly, has affixed and removed a small outboard motor. Usually painted, most hulls boast bright primary colors: reds, yellows, or blues. A census of Puerto Rican fishers found that over 85 percent of fishing boats were under twenty-two feet in length and motorized, most commonly by a twenty-five-horsepower motor (Collazo and Calderón 1988). In some places, these boats may sit by themselves, relatively isolated, seemingly abandoned to the tides and the corrosive effects of prevailing winds. In others, they may be crowded together beside a cluster of buildings, fish-cleaning tables, storage lockers, piers, or other facilities utilized by communities of fishers. Boat-launching locations are differentiated, too, by the presence or absence of recreational crafts; the time of day the location is active; or the extent to which the launching site is also used as a landing and marketing center, where fish are cleaned, frozen or cooked, and sold. Conflicts over space and problems with gentrification are also distinguishing factors. In Puerto Rico, as in other locations, commercial fishers clash with proponents of the tourist industry and other recreational interests over directions of growth along the coast (Martínez and Valdés Pizzini 1996; Griffith 1999).

These variations reflect the diverse ways that Puerto Rican fishers exploit the marine environment immediately surrounding their island and the waters of the Caribbean for a radius of hundreds of miles. Some additional differences among fishing households and fishing communities concern their use of different gear types, or their custom of fishing in shallow or deep water, or the seemingly infinite multiple species that they fish and multiple types of gear combinations that they use. Finer distinctions are also made with respect to household members' participation in fishing enterprises; divisions of labor based on gender, skill, opportunity, and age; and the political behaviors of fishers as they attempt to gain access to varieties of gear, fishing grounds, launching sites, support facilities, and markets.

These lists of distinctions can be added to and revised, more or less continually, as one moves from point to point across Puerto Rican space and time. What at first glance appears to be a quaint, homogeneous, peasant lifestyle based on a time-honored Caribbean tradition is, in fact, a complex and internally differentiated economic, therapeutic, and leisure activity. Fishers may be divided or united by heated political struggle. They have challenged *mayordomos* (overseers) in sugarcane fields, the U.S. Navy on the coast of Vieques, the DNR over access to fishing territories, straw bosses in the tomato fields of New Jersey, and countless U.S. government agencies and British courts of law. They come from densely populated urban areas; from the edges of salt ponds; from wooden shacks in the midst of drought-stricken, thorny trees that look as if they were transplanted from an African plain; and from smooth and cool concrete hilltop homes with vistas as lovely as any in the world.

Yet all Puerto Rican fishers fish under similar environmental, social, and cultural conditions; all share specific constraints, opportunities, and understandings of markets, of the rules of crew recruitment and assembly, and of the typical ways to divide up the fruits of fishing. The first of these constraints and opportunities derive from the sea itself, or from what ethnographers usually call the physical setting.

## FRUITS FROM A DESERT: PHYSICAL FEATURES OF PUERTO RICO'S COAST AND SURROUNDING WATERS

The several islands that Puerto Rico comprises lie in the chain of the largest Caribbean islands of Cuba and Hispaniola, between the Virgin Islands and the Dominican Republic. Puerto Rico is situated so prominently in the sea-lanes that ships approaching the Caribbean can be seen from the mountains that overlook its eastern shore. These larger islands of the Caribbean are called the Greater Antilles, and those of the eastern Caribbean are called the Lesser Antilles. Both sets of islands are surrounded by waters that marine biologists may consider barren in comparison to fishing grounds along the coasts of the Atlantic Ocean and the Gulf of Mexico. The clear, blue beauty that attracts tourists to the Caribbean Sea is a reflection of the low levels of nutrients suspended in a reduced insular platform. With the exception of a few minor rivers, insular Caribbean waters lack extensive nutrient-laden, land-based water flows; long stretches of shallow water; and upwellings (Idyll 1972; Nweihed 1983). Caribbean waters are known as oligotrophic: low in nutrients, with generally low biomass yet with a rich biodiversity of fauna and flora, epitomized by the organisms that are associated with coral reefs, sea grass beds, and mangroves on the narrow confines around insular platforms. The diversity of species cannot, however, compare to the large biomass that characterizes continental areas. The great flumes that emerge from the Chesapeake Bay, the Mississippi River, or the Hudson River Bight have no Caribbean counterparts. Prevailing currents bypass most of the waters adjacent to the islands in two wide arcs that follow the coasts of North, Central, and South America, spiraling around the Caribbean Sea without significantly enriching its waters.

This environmental context deprives the life of the Caribbean Sea of even the unequaled spewing of the Amazon rainforest's richly nutritious death and decay. Consequently, compared to other regions that support large fisheries, such as those of the Pacific Northwest or the Gulf of Maine, the waters of the Caribbean contain low levels of zooplankton and phytoplankton (Griffith and Dyer 1996). Absence of large shellfish or finfish populations comparable to those of New England, narrow shelves, strong trade winds, a hurricane season (from June to October), and seasonal fluctuations in the quality and quantity of fish stocks have restrained the formation of technologically sophisticated Caribbean fishing fleets. Yet in part because of Puerto Rico's special status as a U.S. territory, it is somewhat of an exception to this rule; from its western shores hails a commercial fishing fleet as versatile and sophisticated as the fleets of Chesapeake Bay watermen. These fishers glean groupers and snappers from the deep waters of the Mona Passage, and divers who hail from these and other ports use scuba tanks and spears to make the highly selective catches that are sold in fish markets, out of houses, and along the street. As Figure 3.1 shows, like fisheries elsewhere, the fish landings vary from year to year.

Despite the region's low levels of nutrients, a wide variety of fish and shellfish species are available both seasonally and throughout the year, including those that inhabit deep-water and shallow reefs and near-shore, off-shore, and surface environments. Around the reefs, for example, at least nine varieties of snapper and six varieties of grouper swim with trunkfish, barracuda, porgies, hogfish, and triggerfish. Seasonally, during the winter months, mackerel, dolphin, tuna, jacks, and blue marlin feed near the surface. Closer to shore lurk tarpon, squid, or snook, while on the ocean floor, crabs, octopus, lobster, whelk, and conch inch along. A small pamphlet published by CODREMAR—the government organization

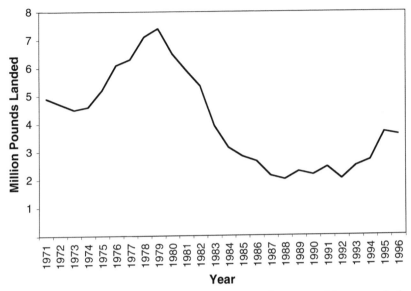

FIGURE 3.1.    Total landings by year, 1971–1996, for Puerto Rico as a whole.

that monitors fishing activity on the island—lists the scientific and common names of Puerto Rican fishes, including 262 saltwater fish species (CODREMAR 1987). As in all fisheries, only a small percentage of these species are targeted for human consumption or sale; others are used or sold as bait fish, and still others are swept back into the sea as components of the complex ecosystem that sustains more desirable and valuable species. Witnessing an hour of beach seining, one might perceive that Puerto Rican fishers expend their energy for little reason other than to return to the sea some of these less desirable, smaller species. Griffith made note of just such a scene in his August 5, 1988, field notes:

> Rincón—This morning we watched eight men work a *chinchorro* (beach seine), surrounding about one hundred yards of water and hauling in a catch of *anchoa* (a small bait fish similar to anchovies). It took one hour to position the net, with one man in a rowboat circling the edge, checking for snags, and shouting orders to the others. Once he had completed the check, he gave the order to pull. The men pulled.

It was obvious that the fishers were having a good time, performing for the crowd that had gathered to watch. Yet after they hauled in the *anchoa,* they seemed disappointed with the catch. Many of the fish they had caught they raked back into the sea with their hands; one nearby fisher encouraged a porcupine fish back into the water. Then they selected individual fish from the catch to keep, most likely for use as bait fish. They seemed to be practicing conservation measures, expending their own energy to return those fish that escaped from the net and could not have made it back into the water on their own.

That Puerto Rican fishers return undesired fish to the sea is just a single reflection of the extent of their love for the sea and its gifts and their effort to pay the sea a form of homage, disturbing as little as possible, as they pass into and out of its life.

Most of the disturbance by fishing takes place over the insular shelf, the 2,335.8-square-mile (583,950-hectare) platform on which Puerto Rico sits (Collazo and Calderón 1988). Along the western shore of the island runs the Mona Passage, famous not only as a source of deep-water, reef-dwelling species but also known as the shark-infested body of rough water across which Dominicans travel to enter Puerto Rico as illegal immigrants. Smugglers use two islands in the passage—La Mona and Desecheo—as stopping-off points before they approach the main island. Called Canal de La Mona by local fishers, the Mona Passage is also a favorite fishing location among anglers from as far away as Ponce, on the South Coast, and Arecibo, on the North Coast. It is also one of the primary fishing grounds for the complex and highly differentiated West Coast fisheries based in Puerto Real, Combate, Boquerón, and Aguadilla.

Containing islands; deep water; rocky stretches of bottom; and shallower, in-shore, muddy, and sandy bottom areas that are easily accessible in small vessels, Canal de La Mona's fishery yields prized snappers; sea basses or groupers; trunkfish; and during certain seasons, the pelagic species of king mackerel, jacks, tuna, swordfish, sailfish, and marlin. The sharks that pass through La Mona are being harvested more and more as markets for shark meat expand. Closer to shore, in shallower water, in Canal de la Mona—as in the near-shore environment all around the main island of Puerto Rico—fishers catch whelk, conch, spiny lobster, octopus, and squid. This traditional space, with a long history of guano extraction, agriculture, hunting, and local fishing, has recently been invaded by sportfishers, recreational divers, eco-tourists, and long-liners near the island. This has increased the number of people who use the passage and has varied the opportunity structure for the passage's traditional users. For example, several fishers from Mayagüez to Puerto Real, with long family traditions of seafaring, now earn portions of their livings by ferrying hunters and tourists to the island of La Mona.

The variety of depths, bottoms, and species of Canal de La Mona is unique among all the waters immediately adjacent to Puerto Rico. It is not surprising that the passage supports the largest, most productive, and most technologically sophisticated fishery on the island; consistently, these waters yield more fish per hectare than other areas in Puerto Rico and account for nearly half of the fisheries' landings. Because this fishery, which is unique in its development, constitutes one logical conclusion for those who pursue fishing as a full-time enterprise and who aspire to construct, from artisanal roots, a capitalist business, the fishery is discussed in some detail at the end of this chapter.

Table 3.1 reflects physical differences among the coasts—including the extensive shelf area of the East Coast and the narrow strip of shelf along the North Coast—as well as the distribution of fishing efforts in the surrounding waters. Although the waters to the east, around the two large island municipalities of

TABLE 3.1. Puerto Rican Fishing Grounds by Size and Productivity

| Coast | Shelf Area (Hectares) | Landings (%) | Productivity (Lbs./Hectare) |
| --- | --- | --- | --- |
| Western (La Mona) | 124,347 | 40.0–48.8 | 24.7 |
| Southern | 123,660 | 25.5–30.0 | 13.0 |
| Northern | 66,639 | 9.4–13.0 | 8.8 |
| Eastern | 269,304 | 16.3–17.0 | 3.8 |

*Sources:* Collazo y Calderón 1988:6; Jean-Baptiste 1999:1–2; Pérez 2000:9–11.

Vieques and Culebra, have the most extensive stretches of shallow water suitable for trap fishing, they suffer lower productivity levels than the rest of the island.

Less than a mile north of the island, the blue of the water deepens where the bottom drops precipitously into the Puerto Rican Trench. Although this is a favorite spot among sportfishers who seek big game marlin, sailfish, and swordfish, the trench itself has not been a fertile fishing ground for small-scale Puerto Rican fishers. Its more than one thousand–fathom depth prevents the development of substantial populations of bottom-dwelling reef fish and causes rough, choppy seas. The North Coast is also the most heavily industrialized and highly developed for tourists. Fishers who live in port cities along the northern coast complain about water quality and pollution issues more than do fishers along other parts of the coast, and those who are located near the western or eastern parts of the island routinely fish to the west or east of their villages.

From San Juan to the eastern municipality of Fajardo, much of the coast has been fully or partially gentrified, developed for fairly wealthy seasonal and permanent residents and tourists. Coastal gentrification is occurring rapidly throughout much of the West Coast and portions of the South as well (Griffith et al. 1988; Valdés Pizzini 1990b). Between San Juan and Arecibo, west of the metropolitan area, lies the most densely populated and industrialized stretch of the entire coast. Fishing to the north of the island is further limited by features of the coastline itself: With the exception of San Juan, Puerto Rico's North Coast has few natural bays and sheltered inlets for launching crafts over rough surfs. In Jobos, for example, just east of Arecibo, fishers must skirt the edge of the shore where curving rock tames a small channel of calm, allowing them to launch their crafts between waves. They time their departures and returns to the roll and surge of the surf.

## PUERTO RICAN FISHING AS DOMESTIC PRODUCTION

We have already emphasized the domestic- or home-production nature of Puerto Rican fishing. Worldwide, most small-scale fishing enterprises are family operations, organically tied to the

growth, maintenance, and reproduction of the family unit
(Doeringer, Moss, and Terkla 1986; Durrenberger 1992, 1995; Griffith 1999; Griffith and Dyer 1996; McGoodwin 1990; Pérez 2000).
In settings that range from underdeveloped countries to the most
advanced capitalist nations on earth, the logic of small-scale fishing usually follows the biography of households, adding, expanding, or subtracting fishing practices and gear to the corpus of fishing activities as children age, crises develop between fathers and
sons, or relationships among households tighten around the pursuit and capture of marine organisms. As such, fishing activities
tend to be born, grow out of, and continually influence the spaces
and phases of the family's life cycle and reproduction.

These spaces include the fisher's house; yard; and, usually, fishing workshop or work space. The work space is either unattached,
part of, or close by the space of the house and yard and does not
constitute a separate, distinct sphere of activity. The following
description of a lot in Guayama demonstrates the symbiosis
between house, yard, and fishing work space:

Ruperto Correa's yard takes up between a quarter and a half
acre across the road from a narrow strip of mangroves that
borders a shallow inland bay in Pozuelo, Guayama. Two
structures sit more or less equidistant from the road: a small
wooden house, around three hundred square feet in size, and
an even smaller concrete-block structure that Ruperto uses
as a fish market. The concrete-block structure is covered
with stucco, painted a light blue, and decorated with the
words "Pescadería Correa" ("Correa Fish Market") in large
black letters. Thick iron bars protect the single doorway,
behind which stand tables where Ruperto cleans his own
catch and that of others who sell to him.

Beyond these two structures, about fifteen yards from the
rear of his lot, begins the clutter of industry that characterizes most Puerto Rican fishers' work spaces. A pavilion with
concrete flooring and a corrugated zinc roof occupies its center. New lengths of lumber support the roof. Beneath the
tent-like structure is a square workbench that reaches to
Ruperto's waist. Ruperto, a large man, is not tall, but he is

thick across the belly and chest, and his size and strength derive from hours of bending rebar into the shapes of fish traps at this workbench. He works methodically, systematically, obviously having performed this task many times before. Materials to make the traps, tools and tool chests, and the refuse and debris of trap construction litter the concrete around the workbench. To the right of the slab, an automobile engine hangs from a tripod; Ruperto allows a cousin who fishes with him to use his tools and his yard to work on the engine. Behind the concrete slab is a small corral where Ruperto raises pigs, and alongside the west fencerow of the lot is a home garden.

At first glance, Ruperto's house and yard seem to be simple manifestations of his reliance on the sea; yet two areas within this work space are telling components of Ruperto's place in the local fishing community. First is the structure out of which Ruperto and his family sell fish. Many Puerto Rican fishers themselves sell fish, either by advertising at the gate of their homes or simply by selling it in the street, but relatively few build structures like Ruperto's. The number of fishers who sell their catch has increased in the past decade or so (Pérez 2000). Ruperto's setup depends on (1) the availability of other family members who can staff the fish market while Ruperto fishes and (2) the specific arrangements he can make with his fishing partners concerning the disposition of the catch. Typically, fishers work on share systems, with one share of the catch going to the vessel for trip expenses and a share apiece to each of the members of the crew involved in the catch. However, variations on this arrangement are not uncommon, particularly among fishers who own fish markets, seafood pastry businesses, or restaurants. In cases such as Ruperto's, he negotiates with his crew for the entire catch, paying them in wages instead of in shares. The size and character of Ruperto's fish market also reflect his withdrawal from the fishing association in his area and support the observation that his fishing community has been experiencing political difficulties. Ruperto, in short, competes with the market that is run by the fishing association, thus undermining the association's general function as a locus of political action among fisheries in Ruperto's area.

The second area of particular interest in Ruperto's yard is the tripod from which an engine dangles. The engine belongs not to Ruperto but to his cousin Juan. The use of Ruperto's tools and work space is one of the fringe benefits that Juan receives for working with Ruperto on his vessel, helping to set and pull fish traps. It is difficult to judge whether this is closer to a gift exchange or a more calculated exchange, in which the privilege that Ruperto gives to Juan will simply ensure that Juan will continue to help Ruperto fish. Working together on the vessel and in the yard requires the mutual trust that typifies gift exchange; yet Juan's behavior in the community at large suggests that certain social divisions separate him from his cousin: He is active in a fishing association to which Ruperto gave up his membership after he became aware that he could not use it to consolidate his fish dealership in the neighborhood.

The tension between family and fishing association that was reflected in the relationship between Juan and Ruperto also reflects the ways that work spaces like Ruperto's supplement, complement, or completely substitute more communal work spaces of fishing. These include *villas pesqueras;* the unaffiliated launching, landing, and marketing centers; and other fishing spaces that draw fishing household members into social interaction and cultural enactment. In areas where the coast is becoming gentrified, developed for tourism, communal work spaces of fishers may become disputed territories, targeted for either destruction or major change as developers cite them as unsightly and indecorous. Similar disputes, which have occurred in Florida, New England, California, and the Mid-Atlantic, are discussed further in Chapter 7 (Griffith 1999; Griffith and Dyer 1996; Meltzoff 1988). Whether or not work spaces are targeted for destruction, the intermingling or separation of home and communal work spaces suggests varying degrees of interdependence between, on one hand, family and household and, on the other hand, the wider community of fishing households. What factors might affect this, by either drawing members of these households into a wider society of fishing or chasing them back into the family circle? In the answers to this question lie some of the essential internal characteristics of Puerto Rican fishing.

## VILLAS PESQUERAS AND THE STATE
## PRESENCE IN PUERTO RICAN FISHERIES

As our observations about relations between Juan and Ruperto imply, one of the principal institutions that mediates the movement between fishing household and fishing community is the fishing association, more commonly known as the *villa pesquera.* On the surface, fishing associations constitute the principal vehicle for state intervention into Puerto Rico's fisheries. At the same time, fishing associations provide a basis, a social and technical infrastructure, for the development of politicized fisheries, with their own internal hierarchies, structures, and relations of dependence. These political features of *villas pesqueras* draw some fishers to and repel other fishers from the associations. Our use of the word "repel" is not an exaggeration in reference to these associations. Many of the fishers we interviewed responded rather vehemently to our questions about their membership in a *villa pesquera.* They made comments like the following:

> No, I don't belong to a fishing association because it's a waste of time and money. There isn't any cooperation, just a bunch of gossips and drunks.

Or, they might say, somewhat more metaphorically:

> I believe the association is, in theory, a good thing, but it is filthy that they help those who aren't fishers but men who stuff themselves on *chavos* [money] and nail themselves to the cross [take government assistance unjustly, merely pretending to represent fishers and the downtrodden]. They leave the organization as rich men and leave the fishers carrying the devil, without leaders.

Throughout the Caribbean, fisher cooperatives and similar institutions, such as the *villas pesqueras,* have been encouraged by governments, fishery managers, and planners (Comitas 1974). As independent producers who operate within the economic boundaries of households and communities, or at least look independent to most outside observers, fishers are viewed as individualistic, often isolated people whose individualism has been posited as a key psychological trait (LiPuma and Meltzoff 1997; McGoodwin 1990; Poggie 1980:21). By contrast, Durrenberger (1995) argues, persuasively, that the apparent individualism of fishers is due less to any psychological predisposition than to historical trends that have undermined the abilities of fishers to organize in the same way that capital has undermined the strength of labor unions. Historically, the potential for fisheries development (measured in higher landings, better technology, and higher ex-vessel prices per catch) lay in organizing these independent producers into cooperatives to circumvent the control of local dealers and merchants (Pollnac 1981:240). It follows that this would also facilitate the sharing or pooling of gear and vessels for the benefit of the group, rather than for the benefit of one or two individuals.

In Puerto Rico, following the U.S. appropriation of the island, the state tried several times unsuccessfully to develop cooperatives for the development of the island's fisheries. In the early 1980s, the government agency CODREMAR initiated a program to develop Puerto Rican fisheries. Working in conjunction with various commonwealth and federal agencies, CODREMAR has built marketing, landing, storage, and meeting facilities for fishers all around the island. Establishing the fishing associations as gathering places for fishers, CODREMAR has also partially centralized the collection of data on landings and other aspects of the fisheries to increase its ability to manage Puerto Rico's marine resources. In theory, these associations allow fishers to organize, centralize, and stabilize their markets; store equipment; and gain access to credit and training. In practice, their impacts vary from place to place: Some associations have become totally defunct, while others have become powerful political and economic forces in their communities. The success or failure of these associations is the result of a variety of circumstances, including the quality of the marine resources available, the fishers' existing technological capacities, and the extent of alternative employment in the area.

Political participation of fishers is also linked to the commoditization of the coast itself. With the demise of sugar and other large-scale agricultural production, uses of the coastal zone have shifted from productive to leisure-oriented activities and infrastructure, a process that is becoming increasingly common throughout the world. Public and private demands on coastal property have threatened coastal communities all around the island. Residents of fishing neighborhoods, many with few or no other economic and social alternatives, find themselves in the political predicament of rejecting development projects that attempt to transform their way of life. Opposition to coastal development projects—from marinas, to resorts, to port and harbor development—often underlies alliances among diverse political and environmental organizations and pits commercial fishers against sport-fishers, harbor masters, private citizens, and hotel and restaurant interests (Griffith 1999; Griffith and Dyer 1996; Meltzoff 1988).

In contrast to cooperatives, fishers who belong to *villas pesqueras* usually own their boats and equipment and are responsible for their production and revenues. That is, they remain petty-commodity producers, owning their own means of production rather than fishing for a fish house or fish dealer, as is common in many fisheries (Durrenberger 1995; Griffith 1999; LiPuma and Meltzoff 1997 Maril 1995; McGoodwin 1990). Since fishing associations must be incorporated into the governmental apparatus in order to function, individuals chosen for the board of directors must have special capabilities. Thus, they are selected from among the most experienced and successful fishers, many of whom also are viewed as social and moral pillars of their communities.

Association presidents usually have some experience in lobbying or in political endeavors. They tend to be well educated or have great experience in leadership, have vast knowledge of bureaucratic procedures, and have contacts in the local or national government. Presidents tend to be bilingual, often as a result of lengthy experience as agricultural and factory workers in the United States. Although they are registered as fishers and have licenses, many also hold other jobs and fish only in their spare time; at the same time, they are full-time specialists in the old art of politics.

Presidents must have the proper rhetorical skills to be able to defend fishers' interests. They must learn the rules of the political game and apply them. Hence, they must be able to use and manipulate the media; lawyers; political parties; and economic opportunities in the local, national, and federal government. Although these may seem to be the ideal characteristics, in fact they describe members of the board of directors of several associations (Gutiérrez Sánchez, Valdés Pizzini, and McCay 1986).

To be successful, associations must maintain political neutrality (or its facade), accepting fishers from the three main political parties and ideologies. Their members, especially their board of directors, must suppress their political inclinations, since their primordial loyalties are not with the party but with the fishers and their fundamental class and community interests. This is especially true for Puerto Rico, a highly politicized society where political affiliation is a sensitive matter, often dividing communities and families. The three political parties differentiate themselves from one another particularly over the question of the island's status vis-à-vis the United States: The New Progressive Party (NPP) is pro-statehood, the Popular Democratic Party (PDP) is pro-commonwealth, and the Puerto Rican Independence Party (PIP) is pro-independence. Given Puerto Rico's lengthy ambivalent relationship with the United States, the issue of its status can be particularly divisive. Thus, an association that fails to avoid a political alliance with or annexation to a party takes the risk of alienating a large segment of actual or potential members.

Presidents who successfully conceal their political interests are also able to manipulate the parties in power, doubling their political capital. To the extent that they are capable of concealing their political party affiliation and are willing to present the association as a unitary block, the board of directors and the president will be in a better position to achieve their goals. Yet effective political leadership in the *villas* is not always benign. At times, fishing associations divide communities by, for example, instigating disputes between powerful fishers and fish vendors, ignoring the views of some community members, or using public facilities and funds for private ends. As a result, membership in an association is an affiliation that is neither universal among

Puerto Rican fishers nor, once established, lifelong. In our sample of 102 households, for instance, fishers were divided roughly in thirds with regard to association membership, with 35.8 percent never having joined an association, 33.9 percent having had and then given up membership in an association, and the remaining fishers currently having membership in an association.

Fishing association membership indicates, at the most basic level, a relationship with an institution that sometimes serves as a marketing outlet for one's catch. At the other extreme, membership in an association entails involvement in a wide variety of business and friendly relationships with other fishers of the community—in particular, the political struggles of fishers, including struggles for or against management or other initiatives of the DNR, the U.S. Navy, or the Caribbean Fisheries Management Council.

The effectiveness of fishing associations as either marketing or political institutions is highly variable from community to community around the island. Those who belong to associations join for a variety of reasons but primarily to sell fish and, often more important, to get access to government aid that sometimes accompanies association membership. Similar to the dynamics of the *gumsa-gumlao* described by Edmund Leach (1964), associations gel into powerful local organizations when leadership and action are needed and tend to disband when there are no threats to fishing ways of life or to the cultural space and time of fishing.

During political disputes, *villas pesqueras* are usually the most active organizations in mobilizing fishers to defend their interests, and membership during disputes typically swells as other associations come to the aid of those involved in the struggle. Valdés Pizzini (1990b; see also Chapter 7) elaborates on a case in which effective association intervention during a proposed marine sanctuary—a development that would have restricted fishing to West Coast fishers—succeeded in opposing the DNR. The use of associations primarily as tools for political organization and mobilization is reflected in the reasons that fishers offer for not seeking membership to associations: They comment either that the association in their area is highly disorganized or that no association exists. While a disorganized association or no association may exist during stable periods, during crisis periods, as when fisheries managers threaten to restrict fishing, we have seen associations from other areas move into regions to aid in the mobilization and organization of fishers. Further, the existence of a powerful or politically active association in a community indicates a dedication to fishing that runs deeper than individuals or households, involving entire coastal communities.

## ACROSS THE ISLANDS: VARIATION WITHIN THE FISHERIES IN PUERTO RICO

With the exception of long-line and tuna fisheries and a small group of fishers based in Puerto Real, fishing throughout Puerto Rico and most of the Caribbean remains primarily artisanal in nature: labor intensive, small-scale, and technologically simple. Yet to characterize Puerto Rican fishing in a way that invokes images of simplicity is to contribute to a bucolic, often misleading image of tropical fisheries everywhere (see, for example, McGoodwin 1990): sails, canoes, outriggers, oars, and pitch or small fishing vessels made from steamed wooden planks or carved from the charred hearts of mahogany trees. From images such as these, how difficult is it to imagine a sleepy paradise of coconut palms, sensuous women, barefoot men diving for pearls, grass skirts, and grass huts?

From technological simplicity, too often, we assume simplicity of life, an image that many of the promotional materials for Caribbean tourism strive to convey. We run the risk of being seduced by apparent simplicity and the consequent risk of missing the underlying intricacies and complexities of social and cultural processes. In a similar vein, Conrad Kottack (1992) struggled self-consciously against the portrayal of a fishing village in Brazil as overly simple, a paradise targeted for assault. Puerto Rico's fisheries appear to be and in many cases are technologically simple; yet they depend on a variety of intricate relationships within households and networks among households in coastal communities.

Like most lifelong fishers, Puerto Rican fishers base their technical achievements on long apprenticeships at sea and on detailed understandings of the characteristics of the various bottom structures of the Caribbean sea, currents, seasonal and regional variations, markets, and the region's wide variety of species of fish. Their selections of gear and target species, varying through the year, depend as well on their home communities, the characteristics of potential fishing partners or crews in those communities, and the character of other jobs they may hold. Against this background, an appreciation of the varieties of fishing gear used by Puerto Rico's fishers is possible.

Four productive strategies for catching fish and shellfish lie at the basis of all Puerto Rican fisheries: nets, traps, hooks and lines, and diving equipment. We conceive of these as the four central or primary technologies, which are usually combined with others yet are inevitably supported by a variety of materials and markets and by relationships within and among fishing households. We characterize Puerto Rican fisheries, then, as fisheries based on primary types of gear yet supported by secondary types of gear.

*Net-Based Fisheries*

Puerto Rican fishers use three types of nets: trammel nets, gill nets, and beach seines. Fishers routinely use other, smaller nets—such as cast nets—as well, but these tend to be used for catching bait and other supporting purposes rather than as primary gear. Of the three primary-gear nets, only the beach seine, called a *chinchorro*, captures fish by movement through the water. Both trammel nets and gill nets hang in the water column. Trammel nets, locally known as *mallorquín*, consist of three nets, usually made of cloth twine, which are bound together like three curtains and entrap fish between them. Sandwiched between two large-mesh nets is a smaller-mesh net that is twice the width of the two outer nets. As a result of this size difference, the inner net floats around with a good deal of slack, improving the apparatus's ability to entangle fish.

Gill nets, usually made of monofilament line, consist of a single curtain of netting. Puerto Rican fishers set gill nets at various depths, based on the kinds of fish targeted and their estimation of the productivity of different waters. Gill nets' local names change with water depths: The net is known as a *filete* if it is hung high in the water, near the surface, where it is likely to catch pelagic species such as mackerel; it is called a *trasmallo* if it is hung on or near the bottom, where it is more likely to snare demersal species such as snapper. Like trammel nets, gill nets have floats along their tops and weighted rope strung through the mesh along their bottoms.

The increase in popularity of gill and trammel nets among fishers in recent years has stimulated several disputes. Many who are concerned about fish stocks and marine ecosystems view nets as a particularly destructive type of gear. Others view nets as physical barriers to navigation and nuisances to recreational activities, including recreational fishing. These views are not restricted to Puerto Rico. Bans against nets have threatened fishing livelihoods in Florida, and restrictions imposed on groundfishing and shrimping—both of which are net based—have reduced fishing incomes from the Gulf of Maine to the Gulf of Mexico (Durrenberger 1992, 1995; Griffith and Dyer 1996; Maril 1995). Restrictions on nets and net fishing have been instrumental in mobilizing fishers to political action in several social and cultural contexts.

Recently in Puerto Rico, disputes over gill and trammel nets have served to further distinguish commercial from recreational uses of the island's marine resources. Although Puerto Rico's tourist and leisure trades have always relied heavily on commercial fisheries for high-quality seafood, recent disputes over nets threaten to drive a wedge between the commercial and recreational sectors. The case for banning or restricting nets in Puerto Rico has been aided by claims that net fishing is a new fishing technique, without any historical basis for its practitioners' claims to space in the sea. The confusion regarding the historical place of net fishing may derive from the modern materials used to weave nets—the monofilament line mentioned previously—or it may derive from the expansion of net fishing in recent years. Whatever its source, we find scant evidence in the life histories of fishers, in historical information, or in other sources that net fishing is a recent technique.

All archival and primary sources that we have examined suggest that gill nets and trammel nets have long been part of the fishing gear employed on the island, but historically they were not the main instruments used in fishing (see, for example, Jarvis 1932; Nweihed 1983). Instead, traps, beach seines, cast nets, hand lines, and fish weirs were the most common types of gear used on the island during Spanish domination. However, various sources indicate that gill nets were used in rivers, river mouths, lagoons, and streams, as well as in the sea, although fewer fishers used them under Spanish rule than later. Several sources indicated that they were also the target of user conflicts. It was during the twentieth century, and especially after World War II, that the gear became popular. With the motorization of the artisanal fishing fleet in the 1950s, the quantity of gear increased, and gill nets and trammel nets doubled their number during that period. For 1990, government sources indicate a record high number of gill nets (788) and trammel nets (507). In forty years, from 1930 to 1970, the number of nets almost doubled, and from 1970 to 1990 the number tripled (Valdés Pizzini et al. 1996).

Again, the recent growth of net fisheries in Puerto Rico may be responsible for the perception that gill and trammel nets are recent innovations, a portrayal that empowers those who oppose nets to imagine them as unsuited to Puerto Rico's marine environment. Several of our life history interviews suggest, however, that net fishing fits well within family fishing operations as households mature, usually sifted into more complex fishing operations and combined with several other types of gear as a household increases its reliance on fishing. Bonifacio Pantojas demonstrates this turn of events:

> Bonifacio uses nets to fish; yet he dives and fishes with hooks and lines as often as he uses his nets, depending, in part, on the disposition of his catch. His fishing has always reflected both his history of migration between the U.S. mainland and Puerto Rico and developments in his own and related households. His father was a fisherman, but he moved to New York with Bonifacio's brother in the 1960s, while Bonifacio's mother remained in Puerto Rico to run her small seafood restaurant. From the ages of sixteen to nineteen, Bonifacio supplied fish to his mother's restaurant, fishing primarily with a hook and line. At age nineteen, he joined his father and brother in New York, but he returned to Puerto Rico after a year to marry and to take his wife back

to New York with him. They remained in New York only three years, returning to Puerto Rico to fish and to help manage the mother's seafood restaurant. Now, however, Bonifacio used a trammel net that belonged to a fish market owner. Most of the net's catch went to the fish market, but Bonifacio withheld enough to supply his mother's restaurant.

Three years later, Bonifacio was divorced, and he returned to New York. Around the same time, his brother returned to Puerto Rico to take his place helping the mother manage and supply the seafood restaurant. This time Bonifacio remained in New York for eight years, where he married a second time. When he and his second wife returned to Puerto Rico, Bonifacio took a job in a hardware store for a year and a half. At that time, he was divorced again, and he returned to his parents' home to live and fish. This third reentry into the fishing community was accompanied by a partnership with another fisher; they combined all their equipment, including nets, and fished together full time. Bonifacio used his share of the catch to supply his mother's restaurant and other restaurants in the area.

Bonifacio's fishing is intimately connected to his migration, occupational history, changing household circumstances, and ability to negotiate access to markets and gear fishing partnerships. These, in turn, are nested in his family relationships: his mother's position in the fishing community as a buyer and promoter of seafood, his father's history of fishing, and his brother's assistance with the restaurant during Bonifacio's pursuit of other economic and social avenues. Net fishing is nothing new to this family, and it has not been incorporated into the household's livelihood in any way that would suggest new pressures on marine resources.

Somewhat similarly, Nestor Torres uses net fishing as the basis for a smaller household business and at the same time cements ties with his cousin's family:

A fisherman in his late fifties, from the Southwest Coast, Nestor uses a *mallorquín* and gill net as his primary gear, combining deep-water net fishing and beach seining with a hook-and-line rig for deep-water fishing. He net fishes with his cousin early in the morning, from 2:00 A.M. to between 9:00 and 10:00 A.M., exploiting primarily in-shore environments, and divides the catch into thirds: one-third for his cousin; one-third for himself; and one-third for the boat, a seventeen-foot craft with a hand-operated seventy-horsepower motor.

Some of the catch Nestor sells to dealers in Puerto Real; yet he withholds a few species for his wife's family business: She makes seafood pastries with the *chapín* (boxfish) that Nestor catches and sells them to tourists primarily on weekends and sporadically during the week. Despite the fact that he supplies the fish, he refers to this enterprise as his wife's business and considers himself a full-time fisher. The Torreses' twenty-six-year-old son lives with them and cleans beaches for a living.

Nestor's net-fishing skills come from his parents' teaching. Both his mother and father fished while he was growing up, his mother more intensively after the death of his father (who had been married three times and had fathered twenty-six children). It is not common for women in Puerto Rico to fish; yet in Nestor's mother's case—whose fishing became a necessity after the father's death—Nestor reports that she acquired the working capacity of a man.

Again, in this case, net fishing occupies a central position in a diversified household economy that engages husband and wife and takes its inspiration from the fishing practices of Nestor's mother and father, thus building on family tradition in a way common among large, extended fishing families.

### Trap-Based Fisheries

Traps have been central to hunting groups for several millennia. Archaeological evidence has shown that aboriginal peoples throughout the Americas depended on snares, pitfalls, log falls, and other kinds of trapping techniques long before the recorded historical evidence of fish trapping in rivers, lakes, and seas. It should come as little surprise, then, to learn that trap fishing is among the oldest and most common fishing methods in the world. Puerto Rico's earliest inhabitants trapped fish by means of weirs as well as traps. Today some of the most successful fishing operations of the East, South, and Southwest Coasts use traps as the central gear in their fishing operations.

Like nets, traps—which have been portrayed as an overly destructive type of fishing gear—have been the focus of political disputes (Griffith and Maiolo 1989). Trap fishers, who have long feared theft of either their traps or the contents of their traps, sometimes intentionally leave their traps' locations unmarked. Many believe that this practice results in ghost traps—that is, traps that are lost as a result of fishers' forgetfulness or that move in rough weather and continue to fish until they deteriorate. Perceived problems with ghost traps have led to rules that regulate the materials used in the construction of the traps: Specifically, the wires that bind the walls of the trap together must be made of metal that deteriorates relatively quickly, allowing the trap to come apart. Such rules force fishers to maintain and repair their traps more frequently.

Trap construction, repair, and maintenance consume a good deal of a fisher's time as well as a good deal of the living and working space available to the fishing family. The construction of traps can involve short-distance travel to cut stakes for the frames; in other cases, fishers use rebar to give traps their shape and structural integrity. Full-time fishers who use traps as their primary gear report that they spend two to three hours at a time in trap construction and maintenance. Although the number of days per week that they devote to this activity varies by the number of traps they fish, weather conditions, household production schedules, fishing territories, and seasonal factors, trap construction and maintenance loom large in the trap fisher's weekly schedule. Trap fishing in Puerto Rico, as in other parts of the world (see, for example, Acheson 1987; Griffith 1999), involves the use of marked territories that take advantage of local habitats and a long (now fading) tradition of communal use of fishing locations (Jean-Baptiste 1999; Posada et al. 1996).

Although the most productive Puerto Rican fisheries are based in the West and depend on deep-water hook-and-line fishing in the Mona Passage, trap fishing is the foundation of some of the most productive fisheries along the South and East Coast of the main island. Stretches of shallow water off the shores of Ponce, Guánica, Fajardo, the large outer islands of Vieques and Culebra, and other southern and eastern locations encourage trap fishing and diving. The close proximity between trap fishers and divers has generated some hostility between the two groups. It is common to hear trap fishers complain that divers steal from their traps. Suspicion of this theft, combined with the general fear of trap theft previously noted, further encourages the setting of traps without buoys for markers or setting them within sight of a fisher's home or other base of operations. Even fishers who worry little about theft need to set their traps within sight of the shore in order to locate them by their positions relative to visible landmarks. This limits the territorial reach of trap fishers and constrains the expansion of trap-based operations. Perceived problems with traps or actual declines in trap catches from theft may lead trap fishers to experiment with other types of gear. The stories of two fishers illustrate this occurrence: Rafael Vales has begun to move from trap fishing to net fishing because of perceived changes in trap fishing's productivity, and Marcos Olmo switched to nets from traps explicitly because of theft.

> Rafael lives in southwest Puerto Rico, near the tourist town of La Parguera, and works as a crewman on vessels that belong to either his brother or one of his uncles. Although he describes himself as a *proel,* or crewman, he selects the vessels on which he crews with an eye toward natural resources and the different kinds of gear that his kinsmen use. Until eight or nine years ago, he fished primarily with traps, but more recently he has been sifting net fishing into his trap fishing because he perceives natural resource changes that favor the nets over the traps.
>
> On board the vessels, Rafael is more of a partner than a subordinate, a status reinforced and reflected by two reported facts: First, he brings some of his own equipment to the partnership, including a motor and forty traps. Second, he receives a full half of the catch, rather than the typical one-third *proel* share. This he markets himself in the restaurants

of La Parguera. He has also been teaching his three young sons—ages fourteen, eleven, and nine—how to fish and construct and maintain gear. Most of their fishing lessons involve making and using nets instead of traps, which suggests that Rafael is preparing them for net fishing.

Born in 1956, Marcos spent his childhood in Guánica (1956–1966), on the Southwest Coast of Puerto Rico, and in New York (1966–1974). He did not learn to fish as a boy, unlike most Puerto Rican fishers. He learned the craft in 1975, at age nineteen, a year after he returned to the island. His initial fishing experiences involved hooks and lines and cast nets, but—after taking a year away from fishing in his mid-twenties (1980–1981) to work for an airline in San Juan—he included both diving and trap fishing in his fishing enterprise.

Marcos's adoption of scuba tanks and traps as his principal gear was influenced first by his brother, who lived in California and knew how to dive, and second by an arrangement he made with a doctor who owned a resort in Guánica and who was a friend of the man who had taught Marcos how to fish. The doctor bought Marcos twenty-five traps to use to fish for the resort and also hired him to do maintenance work around the resort and to take tourists to nearby islands for picnics and boat rides. The tourist guide job allowed him to make use of his English and his experience in New York.

Slowly Marcos began to assume more tasks and more responsibility around the resort, increasing his work load and his stress and reducing the time he was able to spend scuba diving for fish (a practice he associated with tranquillity and relief from job stress). Eventually, this led to tensions between Marcos and the doctor. As a result, in 1986, he quit the resort job to fish full time for himself, relying predominantly on scuba gear but using traps as well.

After nearly two and a half years, in early 1988, Marcos's traps were stolen, and he switched from trap fishing to net fishing. Today, he uses trammel nets as his principal gear, although he still dives and continues to use the hooks and lines and cast nets that he started out with.

Although both of these cases suggest that net fishing is replacing trap fishing, a conclusion that seems to be confirmed by recent CODREMAR data, trap fishing remains a cornerstone of Puerto Rican fisheries. In fact, in some locations, such as Maunabo and Guayama, it serves as the basis of successful fishing operations that supply household seafood markets. Some fishers reason that trap fishing is a sure way to provide a predictable supply of fish throughout most of the year, whereas other methods can be more sensitive to seasonal changes in conditions of the sea or seasonal habits of certain fish.

### Hook-and-Line-Based Fisheries

Puerto Rico's most productive fishing fleet, based in the West Coast city of Puerto Real, fishes for reef fish primarily with hooks and lines, making daily excursions to the shelf that reaches into the deep waters of the Mona Passage. Most of Puerto Real's landings come from exploitation of the shelf drop-off by means of reel lines operated with electric motors. Fishing reaches depths that range from 125 to 300 fathoms. This type of fishing, targeting mainly snappers and groupers, is also performed in the waters of other Caribbean islands such as Nevis, the Dominican Republic, and Turks and Caicos.

The fishers of Puerto Real constitute one of the Puerto Rican fishing industry's most unique faces, but hook-and-line fishing is common throughout Puerto Rico as either a preferred, principal fishing style or a supplementary technique that fishers use when conditions at sea prevent the use of other types of gear. Fishing with hooks and lines is the least politically contentious technique in Puerto Rico, as in other parts of the Caribbean and the United States, because of its popularity among politically powerful recreational fishers; yet commercial hook-and-line fishing includes several rigs that are distinct from most of the simple rigs used by recreational fishers. For example, although Fernando Cruz is primarily a trap fisher based in the southwest port of Guánica, he fishes with several different hook-and-line rigs while his traps soak, varying rigs from season to season as the species of fish change:

- During the summer months, fishing for surface-feeding species such as mackerel and dorado (dolphin), he uses a trolling rig called a *corrida*, which consists of a long line with several baited hooks, drawn behind a moving vessel.
- From November to April, he fishes primarily for dorado but with a rig called a *vara*, consisting of a rod and reel similar to that used by recreational fishers.
- Finally, from December to February, when he is not using the *vara*, he bottom-fishes for grouper and other reef fish with a rig called a *fondo*, a hand line with one or two weighted hooks, similar to the rigs that Puerto Real fishers use with electric reels.

Using these rigs while his traps soak allows Cruz to engage in a productive enterprise while he guards his gear and his catch. As more fishers fear theft of traps, nets, or the fish from this equipment, they may sift ever more active fishing methods into their operations.

### Diving-Based Fisheries

Somewhat different from the other kinds of fishers of Puerto Rico, divers are usually highly specialized fishers who can target their catch with extreme selectivity. Shallow-water skin diving is an age-old fishing technique through the clear blue waters of the Caribbean, as García Márquez suggests in a brief sketch about a character named Euclides:

> Euclides was about twelve years old, and he was fast and clever and an incessant talker, with an eel's body that could slither through a bull's-eye. The weather had tanned his skin to such a degree that it was impossible to imagine his original color, and this made his big yellow eyes seem more radiant.... Euclides was as good a navigator as he was a diver, and had astonishing knowledge of the character of the sea and the debris in the bay. He could recount in the most unexpected detail the history of each rusting hulk of a boat, he knew the age of each buoy, the origin of every piece of rubbish, the number of links in the chain with which the Spaniards closed off the entrance to the bay. (1992:90)

In contrast to free diving, diving with scuba gear—which requires access to an air compressor—is fairly new to Puerto Rico, no more than thirty or forty years old. Many divers point to the combination of the tranquillity and dangers of diving, and many of the younger fishers we interviewed were attracted by the idea of fishing with scuba gear and spears. Diving provides the additional benefit of meshing well with other stationary fishing techniques, such as setting traps and hanging gill and trammel nets. This is evident from a snapshot of Alex Ángel's operation:

> Alex dives around his nets harpooning fish as the nets soak off the eastern Puerto Rican island coasts of Vieques and Culebra and the nearby U.S. Virgin Islands of St. John and St. Thomas. He dives from May through the end of October, but he sets traps all year round, fishes with hook-and-line rigs during the first five months of the year, and sets his nets from January through August. His boat is a twenty-foot vessel with a seventy-five-horsepower motor, and when he dives, he uses a crew of three: two divers and one person to stay with the boat while the others dive. Under this arrangement, he divides the catch three ways, with the vessel getting between 15 and 20 percent of the catch for trip expenses. Only when he dives, he claims, does his vessel not take a share. He observes that many divers become too specialized, failing to utilize alternative gear. "This is the error of divers," he says. Any one gear type, he claims, is bound to fail some of the time.

Despite the overlap of fishing techniques in Alex's operation, trap fishers continue to suspect divers of stealing from their traps, and tensions between the two groups persist.

The fishers and fishing styles described in this chapter characterize most of Puerto Rico's fisheries. Fisheries that base their enterprise on nets and traps usually do just that: They base the operation on those gear varieties but do not become so specialized that net fishers do not also use traps, hooks and lines, spears, and other gear. This flexibility is common among fishers who operate small vessels and tend to combine fishing with other pursuits, moving between fishing and a variety of occupations throughout their lives and even over the course of a single year

(Durrenberger 1995; Griffith 1999). Although flexibility of this kind in fisheries that base their operation on traps and nets is most common across Puerto Rico, an alternative, more highly specialized fishing fleet has developed on the island's western shore. The grouper/snapper fleet of Puerto Real, discussed at the end of this chapter, is unique for being the island's most productive, most technologically sophisticated, and most heavily dominated by merchant capital and dealer relationships. As such, it is somewhat of an anomaly among Puerto Rican fisheries; yet it demonstrates one logical outcome of fisheries development: the increasingly central role of merchant capital in organizing fishing styles, social relationships within the fishery, species targeted, and other characteristics of the fleet. Before we begin that discussion, however, let us consider two groups that compete with Puerto Rico's small-scale fishers.

## ALTERNATIVES TO THE ARTISANAL: SPORTFISHERS AND FOREIGN INDUSTRIALIZED FLEETS

This book concentrates on Puerto Rican artisanal or small-scale commercial fishers, but two other groups of fishers—sportfishers and foreign industrial fleets—ply the beautiful blue waters of the Caribbean. As in other regions' fisheries, both groups impinge on the livelihoods of artisanal fishers in various ways: Industrialized fleets often give commercial fishers a reputation as environmental rapists that organizations of sportfishers then use in their efforts to privatize, or commoditize, the coastal zone. For example, in New England in the mid-1990s, regulations that cut in half general fishers' days at sea were implemented primarily because of a decline in groundfish landings that was attributed to the large, industrialized roller-net fleets based in Gloucester and New Bedford, Massachusetts. Cod fisheries founded in this region of the country were among the first European business ventures in the New World, and the fleets of Massachusetts—owned primarily by fishers of Italian, Portuguese, and Norwegian ancestry—are among the most sophisticated fishing fleets of the world, fishing as near to home as Georges Banks or as far away as off the coast of the Carolinas (Griffith and Dyer

1996; Kurlansky 1999; Vickers 1994). These fleets are the best, most well-known, and most visible of New England's fisheries, but several other, smaller-scale fishers work the waters along the shore of New England—targeting groundfish and other species—and their activities were also constrained by the days-at-sea regulations. Yet a potentially more devastating consequence of the regulations on small-scale and medium-scale groundfishers was their role in exaggerating the relationships between fishing and the condition of fish stocks. Anyone who paid attention to the vast numbers of government announcements, congressional and fishery management council hearings, and other components of the discourse that attended commercial fishing was bound to reach the conclusion that groundfishers—through rapacious practices—had brought the crisis on themselves.

These sentiments help pave the way for sportfishing organizations and other leisure interests, primarily tourism, to change the conditions of access to, and uses of, marine resources. Often this involves revising coastal biographies, laying the groundwork for converting the coast and the surrounding waters from a place of work into a place of leisure (see Griffith 1999:chap. 7). Students of gift exchange view processes such as these dynamically, understanding that the biographies of things—their social lives and cultured histories—change over time, often in close step with changing power relationships. As their biographies change, so change their values, and along the coasts of the Americas, with the shift from commercial to recreational uses of marine resources, marine resources themselves have assumed new values. Even when sportfishers target fish for food, included in the value of these fish are several aesthetically pleasing activities and environments associated with pursuing, catching, and consuming them. Many would argue that these fish have become fetishized or that they have been—in Igor Kopytoff's words— "singularized," like a slave being considered part of the family: In short, they become, through reification, too valuable to treat merely as foodstuffs (1989). Once in Puerto Rico, we saw a street vender advertising *McNuggets de marlin* (marlin McNuggets). This struck us a funny precisely because the growth of sportfishing along Puerto Rico's coast was just then advancing to a point where the idea of catching a marlin to sell along the street would soon border on blasphemy.

In Mayagüez, a declining tuna-processing sector that for many years served an international tuna fleet for companies such as Starkist and Bumble Bee, today attests to the problems that large, industrialized fishing fleets have suffered over the same time period that coastal tourism and sportfishing have blossomed across the island. These fleets remain a feature of the Puerto Rican commercial fishing profile; yet because of their specialized nature and their close relationship to the tuna canneries of western Puerto Rico, they have been less intrusive in the lives of small-scale fishers than the ubiquitous sportfishers. Organized throughout Puerto Rico into *clubes náuticos*—nautical clubs— Puerto Rico's sportfishers have become a new, powerful force in the fisheries of the region. In studies of sportfishers conducted over the last fifteen years of the twentieth century, we found relationships between sportfishers and commercial fishers to range from cooperative and interdependent, sharing space and advice, to openly hostile (see Chapter 7).

Similar variations in relationships between sportfishers and commercial fishers characterize other popular tourist locations. In the Mid-Atlantic, Griffith (1999) found that professional sportfishers, such as guides and charter boat captains, often shared sentiments, family backgrounds, and class affiliations with small-scale commercial fishers. Yet more organized sportfishing groups commonly perceived and actively characterized commercial fishers as environmental rapists and thieves of the people's fish. In Florida, Suzanne Smith and Michael Jepson (1993) have argued, these characteristics led to net bans, other restrictions on fishing, and feelings of low self-esteem among working watermen. E. P. Durrenberger (1995) found similar attitudes among fishers along the Gulf of Mexico, and David Griffith and Christopher Dyer (1996) found the same among New England groundfishers. Further discussion and more in-depth analysis of relationships between leisure and commercial interests in Puerto Rico, in the context of specific disputes, are presented in Chapter 7.

## AN ALTERNATIVE ROUTE IN PUERTO RICAN FISHERIES DEVELOPMENT: THE CASE OF PUERTO REAL

In contrast with most of the Puerto Rican artisanal fisheries, one fishing community in Puerto Rico is a remarkable representation of a case of differential development, characterized by high levels of technology and capital investments in production and distribution. This is the commercial fishery of Puerto Real, in the municipality of Cabo Rojo, on the island's West Coast (Valdés Pizzini 1985).

Puerto Real is the island's most important fish landing center. The municipality of Cabo Rojo also harbors other landing centers, all close to the eighteen-mile-long insular platform that accommodates a continuum of habitats, including coastal lagoons, sheltered bays, mangrove forests, coral reefs, rocky and sand bottoms, turtle grass flats, algal plains (intersected by reefs), and the shelf drop-off. These habitats produce the largest amount of fish and shellfish biomass throughout the island (Wieler and Suárez-Caabro 1980:7). In fact, Cabo Rojo's landings produce nearly 40 percent of the island's total catch, the highest for any municipality. Puerto Real fishers use the inshore habitats, but most of the important landings come from exploiting the shelf drop-off by using reel lines operated with electric motors. These lines fish at depths that range from 125 to 300 fathoms and target highly prized snappers and groupers.

### The Production Units

Of the thirty-seven active production units (boats, gear, and crew) that hail from Puerto Real, slightly over one-half (54 percent) reel-line fish in the drop-off and around other Caribbean islands. Another one-quarter combine trap fishing with occasional demersal line fishing and trolling in the inshore areas. Fishing with diving equipment, along with occasional hand-line fishing, is becoming increasingly popular. Only 5 percent of the units use fishing nets and lines, although the percentage has been increasing along with the general increase in net fishing around the island (Valdés Pizzini et al. 1996). The government defines fifteen large vessels in Puerto Real (43 percent) as trawlers, but they are used in reel-line fishing rather than in trawling. Relatively high-tech, they are equipped with diesel inboard motors, echo-sound gauges, radios, two to four electric winches for the reels, hydraulic winches, sleeping quarters, and a kitchen. Their average size is thirty-six feet. Twenty-two percent of the seacrafts are modern, similar to the trawlers but smaller, averaging twenty-four feet in length. Fishers refer to both the trawlers and the smaller crafts as *lanchas*. The smaller *lanchas* are used for hauling traps in the inshore areas, and a few have been equipped with electric reels for fishing along the shelf drop-off. Small boats, or skiffs, some produced locally, others made with fiberglass, are the second most popular type of vessel used in the fishery. These small boats, called *yolas*, are used by divers, line and net fishers, and trap fishers. Most *yolas* exceed nineteen feet (69 percent) and forty horsepower (67 percent), and most (73 percent) use a combination of materials in their construction, most commonly fiberglass and metal (Valdés Pizzini 1985; Valdés Pizzini et al. 1996).

*Crew*

Vessels in Puerto Real are operated by approximately 85 fishers, mostly in crews of two individuals but in rare cases three and four fishers. Along with boats and gears, crews are organized as follows:

1. Trawlers and modern boats used for bottom fishing using reel line: These target snappers, groupers, and an incidental by-catch of dolphinfish. Crews are composed of the skipper/fisher and two to three *proeles* or deckhands. Fishing trips range from five to fifteen days, depending on the distance of the fishing sites, which obviously eliminates the possibility for any crew member to work a steady job on land. Crew composition here is the least stable in the fishery. Skippers either own the vessels (fishing them or subletting them to others) or sublet the vessel from their owners. Captains who sublet often change vessels, and most change *proeles* more than once a year. Unlike other types of crew arrangements, the *proeles* come from other communities in the municipality rather than from Puerto Real.

2. Modern boats, fishing smacks, and *yolas* used for trap fishing:
   Traps are usually hauled with a gasoline winch, targeting lobsters and reef fish. The number of traps per vessel ranges from
   20 to 185, with an average of 102 traps per crew. Crews are
   composed of a skipper and a deckhand. Fishing trips take place
   around three days a week for five hours a day. Trap fishing
   involves four types of crew relationships: siblings, fathers and
   sons, in-laws, and friends (non-kin), with sibling and father-
   and-son ties most common (Valdés Pizzini 1985).
3. Small boats or *yolas* equipped with outboard motors and crews
   of one or two fishers: Fishers operating these crafts fish with
   lines and nets along the insular shelf. Teams of divers use
   them as well—most commonly two divers and a pilot—targeting lobster, conch, and reef fish. Fishing trips occur daily.
   Most of the divers are not related to one another.

*History of the Fishery*

The most prominent characteristic of the fishery of Puerto Real
is the existence of private fish houses or *neveras* (literally, "refrig-
erators"), where fish are bought from fishers and sold either
locally or wholesale to restaurants and other buyers. Early in the
growth of the Puerto Real fishery, the ability to conserve fish for
sale at a later date became the most important characteristic of
these operations. Out of seven *neveras*, three have a long life history and four are newer operations. The *neveras*, owned and operated by fish dealers or merchants, also own fishing boats and gear
that they sublet to skippers and crew. Despite their investment
in production, their profits derive largely from the circulation of
commodities: buying and selling fish, gear, equipment, fuel, ice,
frozen imported bait (squid and mackerel) and other supplies that
have both use and exchange value for fishers as well as for the
general public.

Although the scant archival information suggests that, during
the nineteenth century, Puerto Real was an important harbor and
fishing center, there is no concrete evidence of commercial fishing, except for sales of turtle shells to San Juan, probably for
export (Valdés Pizzini 1985). Both in government reports and our
informants' accounts, commercial activities started in the 1930s,

when local fishers began to sell fish in nearby towns. Fish dealers from the municipality moved their operations to Puerto Real
at the onset of the demise of the local maritime commerce and
transportation. Following a common path of peasant differentiation, these merchants established business relationships with
local fishers who were interested in selling their catch by forming partnerships with the dealers (see Roseberry 1976). These relationships often involved credit for fishing expenses and gave merchants some control over the disposition of fishers' catch.
Typically, such relationships encouraged some fishing styles over
others, as merchants demanded the most highly valued and easily marketable fish. Some fishers avoided partnerships with merchants, and by using family, including enlisting their wives into
the management of their fish houses, local fishers established
commercial operations. Profits obtained by the dealers who sold
fish attracted other dealers from the municipality to establish
their operations in Puerto Real.

By investing their capital, by establishing kinship relationships
with the local fishers, through *compadrazgo* and marriage, fish
dealers eventually became embedded in the fishery's social system and quickly assumed dominance in the relations of production. The dealers invested in infrastructure, including piers, ware-
houses, freezers, and satellite vending operations, in nearby cities
and towns. By financing fishing trips, they were able to buy the
fish at a low price. The growing social differentiation between
producers and buyers was consolidated in the 1940s. In that
period, the dealers acquired fishing gear and boats to sublet to the
fishers, which allowed them to buy the catch at lower ex-vessel
prices. By buying a large number of fishing smacks and adapting
small gasoline motors to the boats, the dealers increased the productive capacity of their operations, as well as the volume of fish
landed for sale. Subletting; the financing of fishing production;
price control; control over fishing equipment, fuel, and ice were
established by the dealers as strategies to maintain tight control
of independent producers and semiproletarians while they managed to increase their profits (Blay 1972; Poggie 1979). As indebt-
edness became pervasive, the alternatives for the fishers were
reduced to abandoning the fishery; selling their boats (usually to

the dealer); or having another dealer buy their debt, a process through which their labor became fully commoditized.

As the *neveras* increased the size and quality of their operations, they also recruited people from fishing households as wage laborers to work at the fish houses for regular pay. Through this process, the *nevera* became the school in which fishers—as clerks, truck drivers, and salespersons—learned the skills and strategies of becoming dealers. With experience and earned savings, these laborers invested in pickup trucks and ice boxes and started to peddle fish throughout the island; some were able to buy or build their own *neveras*, entering into competition with their former bosses.

Early in the 1970s, merchants who owned most of the fishing smacks and *neveras* organized the fishery. Thus, these merchants controlled the market and labor relationships while being able to exact surplus value and profits (Dibbs 1967). Merchants consistently invested in new equipment, allowing their production units to increase the catch. In the 1970s, the fishing smacks were motorized with diesel and gasoline engines; this permitted them to fish for snappers and groupers along the shelf drop-off and on the fishing banks of La Mona Passage, between Puerto Rico and the Dominican Republic. The majority of the vessels devoted to trap fishing had mechanical trap haulers (Abgrall 1975). Although fish traps were the primary fishing gear, increases in ex-vessel prices for snappers encouraged a shift from traps to reel-line gear, becoming a strong rival to the traditional trap-fishing methods (Poggie 1980).

Late in that decade, the government initiated various programs by which large boats (such as the trawlers and modern boats described earlier) were sold to the fishers throughout the island, with the idea that the larger the boat, the larger the catch and profits. Unfortunately, inappropriate methods of technology and information transfer, combined with high vessel prices and operating costs, resulted in the program's failure in all of Puerto Rico except for Puerto Real. With the navigation experience derived from merchant marine experience and fishing in La Mona Passage, exploiting the high-valued resources of the snapper/grouper fishery, fishers from Puerto Real were able to use the technology efficiently. The trawlers for reel lines and the smaller boats for traps in a large platform area, with an increase in the number of traps from 20 to 150, could maintain a profitable operation. But not all fishers were successful at this enterprise; some could not pay their debts, and production costs remained high, since merchants controlled the prices for the factors of production. The dealers applied for loans to buy vessels and also acquired them by buying the debts of the fishers.

This decade of transition was marked by an increase in the fishing technology and harbor infrastructure, aided by government programs. This increase was also highlighted by a boost in the landing of snappers and groupers, most of which were destined for an expanding tourist trade. By the end of the twentieth century, after a period of capital concentration and much developed tourism, the Puerto Real fishery was dominated by three dealers who operated *neveras:* All came from fishing families from the settlement or from the municipality, all bought boats and facilities from former dealers from the community, and all deepened their relationship with the community through *compadrazgo* and marriage.

## CONCLUSION

Puerto Real's fishery represents one possible outcome of fishery development in Puerto Rico. Although it is exceptional, the centrality of merchant capital in the organization of the fishery, combined with strategic infusions of capital from the state, are important considerations in many of the behaviors that we observe elsewhere in Puerto Rico. Yet unlike cases elsewhere (see, for example, Sider 1986; Vickers 1994), even as the Puerto Real fishery became dominated by merchant capital, these merchants became part of the fishing community through kinship and friendship ties. The interpenetration of work and family in Puerto Rico's fisheries is particularly crucial as fishers struggle to juggle fishing and wage labor in their household production activities, even as they resist domination by merchant capital or attempt to dominate others through marketing arrangements of their own. Determining how much time to devote to wage labor, family, community, association and the time at sea can consume as much of a fisher's thoughts as his or her attention to the many and varied details of ropes and fuel and lines, turning fishing in Puerto Rico into a delicate balance between stress and reward, hope and disappointment, injury and therapy. We now turn to those very balancing acts.

# 4 Chiripas

## Working-Class Opportunity and Semiproletarianization

WHILE WE WERE TRAINING Puerto Rican field assistants to interview fishers, we recommended that, during several portions of the life history interviews—especially when fishers recalled their work histories—the interviewers might probe for more information. Specifically, we suggested that they ask about *chiripas*, the colloquial term for casual, varied work—what we would call, in colloquial English, odd jobs, though these jobs are not "odd" at all but are central to many working households' strategies. On hearing this term, the assistants laughed and joked with one another that the work of interviewing might well be considered a kind of *chiripa*. It was temporary work, with pay based on numbers of interviews rather than numbers of hours and with work schedules that changed from week to week. Thus, the work of interviewing did resemble the myriad varied jobs of uncertain duration that were taken from time to time by nearly all of the fishers we interviewed, in order to make ends meet.

Those informal jobs that people wedge into their schedules—jobs whose temporal qualities differ, sometimes quite radically, from most work in the formal economy—are common among people whose primary occupational identities derive from seasonally variable endeavors. *Chiripas* fill in during dead times, the times of little or no work and little or no pay—what the workers in the sugarcane fields of Vieques, one of Puerto Rico's outer islands, call *la bruja*. This term probably comes from the fact that this time of reduced household income and reduced consumption affects people the way most witchcraft affects people, causing weakness, tensions within communities, illness, and death. During dead times—whether those of sugarcane workers who await work in the cane, fishers who sit out choppy seas, or peasant farmers who are caught between harvests—people with time on their hands and families to feed seek work and income wherever they can find them. A differentiated economy helps, as do skills that can be converted into crafts or other products or can be sold in a labor market. The necessities of dead times encourage the development of skills—human capital—and promote community developments that further differentiate local economies. This is one reason that highly internally differentiated, if also redundant, informal economic practices develop along the margins of plantations, haciendas, and other seasonal industries.

Proletarianization encourages people to move in the opposite direction. Don Taso's account of work in the cane illustrates a gradual loss of specialized skills and their replacement by standardized production methods that required little special training, in contrast with the long periods of apprenticeship that fishers require to learn their craft. About this Mintz writes:

> The standardization of work meant the need to concentrate all one's energies in the direction of increasing productivity by wage labor and using such productivity as the prime measure of one's economic worth. . . . The company stores came to exercise control over men's use of their time and earnings, and incidentally exposed workers to a greater variety of attainable commodities. The men's feelings of security that came from their personal relationships with superiors diminished, and some other source of emotional (as well as economic) support had to be substituted. . . . These many changes had the effect in part of "proletarianizing" the people of Jauca. (1960:175)

Yet proletarianization is rarely a lineal, unitary process in which class formation proceeds smoothly. Class formation always leaves behind a litter of victims and, inevitably, a handful of beneficiaries. Particularly with global integration, new opportunities develop while others become altered or obsolete. In Puerto Rico, following the U.S. takeover of the island, fishers who operated fish weirs saw the new political climate as a potential source of support for their claim to property rights regarding placement of weirs (Valdés Pizzini 1990a). Further opportunities developed during the transformation of haciendas into capitalist farms, the 1917 confirmation of citizenship, increasing migration opportunities to the U.S. mainland, state investment in fisheries, and the somewhat unique economic development stimulated by 936 tax laws (Duany 2000; Morris 1995).

With each such development, the potential exists for the process of class formation that is underway in one region under one mode of production to become frustrated by opportunities that develop in other regions, in other industries, or within another mode of production. This is similar to what political economists call crises of accumulation: historical junctures at which the rules of appropriation and exploitation change; surplus values created in one place are siphoned away to another; and struggles to gain or recover people's time, energy, allegiance, identity, and ideas ensue.

Puerto Rican fishers routinely take advantage of new employment, income, and consumption opportunities. Fishing, though central, is only one among many sources of household income. The jobs taken in the formal economy tend to fall within a narrow range, at the low end of occupational hierarchies. Many require little or no skill, pay relatively poorly (at or slightly above minimum wage or with great seasonal fluctuations), and often fall into the class of jobs that are hazardous or highly undesirable (Griffith 1993). As a result, Puerto Rican fishers are rarely intimately tied to the jobs they take in the formal economy, either economically or in terms of identifying with a specific occupation. Instead, they draw from fishing the sense of purpose, identity, interest, tranquillity, and other qualities that those of us who are married to our jobs draw from our own work.

Early in our fieldwork, as we began to elicit life histories from fishers, we started to compile a list of jobs that fishers had performed or were familiar with because their spouses, daughters, or sons had performed them at one time in their lives. As we point out elsewhere in this book, fishers need not hold a particular job directly in order to consider it part of their occupational repertoire (that is, part of the corpus of occupations within fishers' skill and knowledge levels, occupations within their reach through training similar to the apprenticeship of fishing, or occupations held by other members of their networks and households).

We listed thirty-four occupations, which range from relatively low-wage, unskilled jobs such as sugarcane cutter to skilled jobs such as teacher or nurse:

| | |
|---|---|
| Cane worker | Owner of fishing equipment |
| Migrant farmworker | Mechanic |
| Fish cleaner/processor | Policeman |
| Taxi driver | Janitor |
| Shop owner | Government employee |
| Maker of fishing equipment | Company employee |
| Animal raiser | Cane cutter |
| Stevedore/longshoreman | School teacher |
| Security guard | Cook |
| Store clerk | Diver |
| Puerto Rican factory worker | Motor repairer |
| Puerto Rican farmworker | Street vendor |
| Construction worker | Farmer of own farm |
| Fisher | Fireman |
| Health technician/nurse | Waiter |
| Merchant marine | U.S. factory worker |
| Boat builder/repairer | Fish market owner |

We wrote the names of each of these occupations on cards, one occupation per card, asked fishers to sort the cards into piles of occupations that they considered similar to one another, and then asked them to tell us about their reasons for constructing each of the piles. This is a simple cognitive technique that most people have little trouble finishing. A bonus to the technique is that the cards provide visual cues to get people to talk about occupations (Johnson and Griffith 1998).

When Puerto Rican fishers sorted the cards, they tended to separate occupations associated with fishing from the other cards. In and of itself, this occurrence is not surprising. Yet it also seemed consistent with the general way that fishers think about work. Fishers characterized occupations less by the unique characteristics of the jobs themselves than by the ways that different occupations fit into broader social contexts. For example, the fact that, among the fishing-related occupations, "fish market owner" was conspicuously absent reflected the common distinction that fishers make between merchants and fishers. As noted at the end of Chapter 3, this is a distinction with deeply meaningful, concrete manifestations. Another typical outcome of the sorting was that all the jobs associated with the government were put in the same pile. These piles included jobs that many of us typically associate with government, such as police and fire department positions; yet they also included construction jobs, presumably because so many of these positions that are open to Puerto Rican fishers come from public works projects. For many, government work differed from jobs in agriculture, and both agricultural and government work differed from work in the service and commercial sectors, such as shopkeeping or clerking. Jobs in all three areas, again, differed from most of the fishing-related work.

Not all fishers sorted the cards in this way. Variability within the group was extensive, a reflection of the simplicity of the method as well as of fishers' work experiences. The generalizations that we draw from these data are, necessarily, seated in other observations of and discussions with Puerto Rican fishers; still, the fishers' comments about the occupations are telling. Few fishers, for example, differentiated between occupations based on hierarchy, skill level, income and advancement potential, or other indicators of class: They were more likely to put together, say, a farmworker and a farmer who owns his or her own farm, because they work in the same space.

Similarities of place account for the pairing of occupations in other ways, such as (1) waiter and cook, (2) shopkeeper and clerk, (3) longshoreman and merchant marine, (4) taxi driver and mechanic, (5) mechanic and motor repairer, and (6) farm owner and farmworker. Pairs such as these not only reflect functional relationships within job sites but also suggest that Puerto Rican fishers do not immediately consider factors such as prestige, independence, and hierarchy in their assessments of occupations.

This is not the same, however, as dismissing these factors as completely unimportant or irrelevant to the experience of working and fishing. Yet we mention the occurrence here because similar research performed by Michael Burton (1972) found that, in the United States, respondents sorted occupations according to exactly these principles of prestige, independence, and hierarchy (along with whether or not the work was performed inside or outside). Burton's informants were college students, however, instead of fishers, which raises either interesting questions about the relationships between class position and views of occupations or mundane questions about the influence of education on the performance of pile-sorting tasks. Do Puerto Rican fishers view occupations fundamentally differently from the ways that young middle- and upper-class women and men of the U.S. mainland view them? Do Puerto Rican fishers pay little attention to the qualitative distinctions between jobs, even when those features are also related to daily occupational dimensions, such as comfort in the work place, or to social divisions based on income, prestige, cultural capital, and access to the means of production? Or is it simply and disappointingly that the technique of the cards works better with college students, who are used to being tested, than with Puerto Rican fishers?

It is likely that the answer lies somewhere in between. Questions of theoretical importance are rarely resolved fully without some reference to methods. Certainly, to examine pile-sorting data independently of fishers' work experiences, whether at sea or in a New Jersey cabbage field, is as ridiculous as trying to determine the work practices of inner-city youth from the U.S. Census.

What, then, is the role of these work experiences? They are markedly different from those that we expect to find among U.S. college students or even U.S. citizens in general. Although Puerto Rican fishers derive their principal occupational identity from fishing, their work histories are rich and diverse, replete with so many different experiences that it becomes difficult to recognize trends, categories, or other features that social scientists like to see.

Nevertheless, the work histories that we elicited reflect changing opportunity structures in Puerto Rico and the U.S. mainland combined with a close association between processes of household formation and work. Typically, after a period of wagework in Puerto Rico, many expand their work experiences by migrating to the U.S. mainland. Julio "Marty" Seba, a fisher from Maunabo (discussed in more detail in Chapter 5), for example, began to work outside of fishing when he took a job in the cane fields for one year. Then he migrated to the U.S. Northeast, where his work experience assumed a trajectory common among new immigrants. Initially, he did farmwork; then he moved into the less-seasonal turkey-processing industry and subsequently worked for three years in construction. In two related studies, we found this progression of occupations—from farmwork to food processing to construction (the last often beginning with landscaping around construction sites)—to be common among Mayan refugees and new immigrant Mexican workers (Griffith 1993; Griffith et al. 1995). Puerto Ricans differ from these groups in that, as U.S. citizens, they are less dependent on labor contractors than are Mayans and other new immigrants. As citizens, Puerto Ricans can and often do invoke state protections when they suffer abuse in labor markets. The several occupational injury claims that we encountered in the field lend some evidence to this fact.

In Seba's case, as in many others, the transition from agricultural work in Puerto Rico to agricultural work in the U.S. Northeast reflected an increased government presence in the organization of migrations to the U.S. mainland—a process that complemented other state-sponsored migrations, such as the 1941–1964 Bracero Program and the 1943–1992 British West Indies Temporary Alien Labor Program (Calavita 1992; Galarza 1964; U.S. Congress 1978). Both of these programs, along with the Puerto Rican presence in agriculture, have since been usurped by various waves of documented and undocumented Mexican, Central American, and West Indian workers. A reduced state presence in agriculture was driven by reduced demand for agricultural labor on the U.S. mainland and in Puerto Rico, along with developments (such as housing schemes called *parcelas*) in the labor supply regions (Griffith et al. 1995). Yet the work histories of Puerto Rican fishers are not always simple reflections of supply and demand. Shrinking opportunities in agriculture, combined with developments in fisher households, produced other common blends of migration, formal and informal work, and fishing in Puerto Rico. Too often we attribute these changes of space, work, and social context to economic calculus, when more fundamental human processes may be at work. The experience of Bonifacio (some of which we profiled in Chapter 3), suggests this:

Bonifacio, who had begun to fish as a young boy, first entered the formal labor force in 1967, at age nineteen, when he followed his father and brother to New York to work in a noodle factory. His mother, then forty-two years old, stayed behind to run her restaurant, which was a small establishment that occupied the floor below the family home. Only a year after taking the job in New York, Bonifacio returned to Puerto Rico to marry and take his wife back to New York. Three years later, he and his wife returned to Puerto Rico. There Bonifacio helped his mother by supplying her restaurant with fish that he caught by diving and by using hook-and-line rigs and nets. Fishing slowed after three years, and Bonifacio, who was having marital problems, again left for New York. At this time, his brother returned to Puerto Rico to take his place assisting their mother in the restaurant business.

Once again in New York, Bonifacio first worked for the New York City Parks Department and then took a job as a shipping clerk for a factory that manufactured plastic purses. He was laid off from the factory job in 1981, but he stayed in New York until 1983. At that time, he returned to the island, married again, and took a job in the hardware store where his new wife worked. By this time, his father had returned to Puerto Rico to fish for the restaurant, but eventually his arthritis prevented him from continuing to work at sea.

Bonifacio's second marriage lasted only two years. He moved back into his parents' home and assumed near complete control of his mother's restaurant as her health began to fail from diabetes. During this period he resumed diving as well as fishing with hook-and-line rigs. Just prior to our interview, he had expanded his fishing operation by buying a boat in exchange for three hundred dollars and a gold chain that he had made by hand.

Bonifacio's migration and work history combine economics and human capital skills with matters of the heart. His decisions to move between Puerto Rico and New York result as much from processes of household formation and dissolution as from economic need. Clearly, the fact that his is a transnational household enables Bonifacio to achieve this dynamic between money and love, and his mix of jobs and work sites, of domestic locations and dislocations cannot be reduced to simple footnotes of grand political and economic developments.

Bonifacio's propensity to mix matters of business with matters of the heart is similar in spirit to many fishers' continual desire to return to the sea as a therapy. This is a point that we emphasize again and again and that we develop fully in Chapter 5. Yet Bonifacio's case also suggests that Puerto Ricans will leave their natal villages and their original labors of love because of developments within the family that have little to do with economic need. Many Puerto Ricans have had and still have the option of migrating to the U.S. mainland under the pretext of economic need in order to escape developments of a more personal nature: crises of family and relationships, various community problems, gossip, shame, humiliation. The transnational character of many, if not most, Puerto Rican households (a transnationality that is unique in that it involves a cultural instead of an international boundary) facilitates this ability to make deep, potentially lasting mistakes in one cultural setting and find refuge in another. The overemphasis on (1) political developments in the stimulation of the origin and development of transnational communities among refugees and (2) economic circumstances and discrepancies among international labor migrants have overshadowed the emotional values of transnationalism (Basch, Glick-Schiller, and Szanton Blanc 1995; compare Griffith 1997). However, our analysis of the trajectories of fishers reveals that most of their movements and decisions came in critical moments of the Puerto Rican economy or the economy of the U.S. mainland, such as during recessions or periods of deindustrialization. Their decisions thus derive from an array of motives that often synchronize well with difficult situations and make moving easier. These people may move, that is, because of a marital problem, but often this occurs also at a time when others around them are moving as well.

We raise this issue because of the similarities between, on one hand, arguments and observations that attend transnationalism and, on the other hand, arguments that enable us to make sense of relations between domestic production and wage labor. Both involve movement, if not across space then across methods of economic calculus and value, and both involve a divided presence: people who spread themselves between two realms of behavior and, as Portes says, who "are often bilingual, move easily between different cultures, frequently maintain homes in two countries, and pursue economic, political and cultural interests that require their presence in both" (1997:812). Semiproletarianization—the incomplete insertion of fishing households into capitalist social structures and power relationships—occurs by means of similar behaviors and similar entanglements of hearth, home, heart, and paycheck. Experiences of several fishers, presented below, flesh this out in different ways.

Bonifacio's experience, which involves migration, encourages the linking of transnationalism with semiproletarianization; yet movement between wagework and fishing occurs within many Puerto Rican households without migration. Some fishers, that is, fish part time while they maintain a full- or part-time job in their home communities. Others combine wagework with seasonal fishing, with a degree of regularity, as was common among sugarcane workers who fished during *la bruja*. Still others engage in the sporadic pattern of working for wages for extended periods, often several years, and then, like Bonifacio, moving back into fishing between these long periods of wagework.

Combinations of these variants are possible as well, with some fishers never fully disengaging from wagework yet managing to sift fishing into their work in different ways throughout their lives. Fishers who never fully divorce themselves from fishing have limited opportunities for migration and generally confine themselves to working part time and fishing in their home communities. This is the experience of Fernando, whose story we told in part in Chapter 3:

> Fernando grew up on Puerto Rico's Southwest Coast and first learned to fish by collecting octopus and small fish and shellfish along the shore. His father, who fishes as well, now serves on Fernando's boat as his crewman, just as, during an earlier period, Fernando served as *proel* on his father's vessel.

Fernando's early wagework included a brief stint as a full-
time local bus driver in 1978, when he was nineteen years
old, a job he quit after an argument with his supervisor. Now
he works part time for the marine sciences department of the
University of Puerto Rico, which maintains a small field
station and laboratory near Fernando's home. This job is
flexible enough that he spends more of his work time fishing
than working for the university. In fact, even while he had
the job as the bus driver, he fished on weekends and holidays.

Fernando is primarily a trap fisher, although he also fishes
with hand lines, varying the rigs and locations through the
year. His operation draws on the labor of all the members of
his household, including his wife and three young children:
two sons, ages twelve and ten, and a daughter, age eight. He
has one hundred traps at any given time, and he must continually work to build new traps and repair old traps, dis-
carding some every eleven or twelve months to what he calls
"the graveyard of traps." His family pitches in with this trap
building and maintenance, as well as with the myriad tasks
associated with boat maintenance, fish cleaning, and other
pursuits. They also help him build and repair boats, a small
subsidiary service that he provides to the community.

Fernando's fishing enterprise involves the active participation of members of two other households: those of his father
and his uncle, who owns a fish market. Further, Fernando is
an active member of the local fishing association, insisting
that the group's support of fishers in the community is similar to that of a labor union. Most important, the association
recently opposed the creation of a marine sanctuary that
would have denied Fernando access to some of his most
prized fishing locations. Not all of Fernando's fishing locations are widely known, however, and he is especially careful
to set his traps in secret when he targets lobster.

One of the more notable features of Fernando's operation is his
depth of commitment to the local fishing community. He participates in the local association as a member and builds and
maintains vessels for other fishers in the community. His fishing

operation requires a high level of meticulous care. He estimates, for example, that it would take him two and a half weeks, working around the clock with his whole family, to make one hundred traps. And he says that each time he soaks his traps—three days per week—he expends three hours in cleaning, boat preparation, and boat and trap maintenance. On top of this, according to a complex annual schedule and a variety of depths and rigs, he fishes with lines while his traps soak (see Chapter 3). In Fernando's words, *"Cada día se hace algo"* ("Every day one makes something [or money]"). Finally, this complexity extends to the social realm, bringing two other households, his father's and uncle's, into his own operation.

Does this complexity require Fernando's constant presence? Does this limit the extent to which he can participate more fully in wagework? Does this kind of devotion to fishing depend on the commitment of an entire household, or even three households, which in turn depend(s) on Fernando's continued attention? These are questions that Fernando's experiences raise, questions that probe into the logic of domestic production as it is influenced by its practitioners' working (or not working) in other occupations. It is worthwhile to add that Fernando's operation and the related fishing enterprises of his father and uncle have secured them a fairly high standard of living. Fernando's house and his uncle's house, which neighbor each other, exude affluence: His uncle's has an attached fish market, and Fernando's, a two-story dwelling, has an attached workshop.

Fernando's flexible participation in the formal economy does not seem to have compromised his fishing operation and may even have facilitated its growth. Is this a characteristic of semi-proletarianization in general? Or is there a quantitative point to be passed, or a qualitative dimension to be fulfilled, before which fishing is merely a supplementary activity of wagework and, hence, in the abstract, a household subsidy to capital? These are central questions in the growing literature on relationships between domestic production and wage labor (Collins and Giménez 1990; Gringeri 1994). We can begin to address them by first probing into the experiences of a few fishers in terms their differences from and similarities to the experiences of Fernando:

Unlike Fernando, Ernesto Díaz Clausel, a forty-year-old line
fisher from Aguadilla, on the Northwest Coast, spent the
early part of his working career outside his natal community: For fourteen years, from ages sixteen to thirty, he
worked in a publishing house in New York City. With the
accumulated savings from this work, he managed to return
to his home in Puerto Rico to build a house and buy a boat, a
motor, and fishing gear. At the same time, he bought a car
and began a taxi service as an independent operator. The
fishing and the taxi business provided a living for him until
he developed arthritis eight years after he returned to the
island. At this time, he returned to New York for a few
months of treatments, which put him in debt. Now he fishes
full time, no longer drives the taxi, and from time to time
takes jobs doing masonry work as an independent contractor,
what he calls construction *chiripas.*

When Ernesto returned from New York the first time, his
two sons—Robert and Alfred, ages five and three—were too
young to help with the fishing. By the time we met him, ten
years later, he described his sons' participation in the operation in the same enthusiastic terms that Fernando used to
describe his family: *"Los hijos,"* he said, *"me ayudan en
todas las actividades de la pesca"* ("My sons help me with
all fishing activities"). And he told us how they helped him
process and prepare the catch for market, repair and construct gear, and net bait.

Also like Fernando, Ernesto went into great detail in his
description of his fishing operation. He fishes almost exclusively with lines, primarily for large pelagic fish such as king
mackerel; tuna; marlin; and dolphinfish, or dorado—all
highly prized market fish. Much of his time is spent experimenting with and developing new lures, which includes
examining the stomach contents of fish to determine the
best lures. To accomplish this task, he must spend time
searching for materials (such as certain kinds of feathers) for
use as lures. Any lures that he purchases he alters according
to the designs that he devises based on the stomach contents
of the fish.

On the water, he continues his experiments by trolling at different depths during different times of the year, working his rigs in the currents in ways that he has learned to associate with different species of fish. He holds his children and his crew to his own high standards of fishing, and if the single *proel* that he recruits from the community to fish with him knows little about fishing, he gives him less than the customary one-third of the catch.

Although Ernesto rivals Fernando in his appreciation for the sea and his obvious enthusiasm over the technical dimensions of his craft, the two part ways in terms of the respective complexity of each enterprise. Fernando's operation is not only technically more complex than Ernesto's, combining traps with line fishing at various depths and with nets; it is also culturally and socially more complex. Ernesto belongs to his community's association, but just barely. He is neither active in nor in agreement with many of the association's activities. He believes, for example, that the association exists for the benefit of a handful of privileged, politically powerful fishers, who are able to store equipment at the association's facilities even if they hold full-time jobs.

Like Fernando's experience, Ernesto's raises questions about the relationships in each community of fishers and the character of each fisher's participation in the formal economy. At one level of analysis, of course, this becomes a frequency question, whose solution lies in statistical tests such as regression. Too often, however, statistical analysis glosses over important variations from normative trends and leaves us with an account in which human agency and creative response become lost in standard deviations and cases are dismissed as outliers. A more fruitful approach is to examine the logical consistencies and contradictions that emerge from individual cases in light of our general knowledge of Puerto Rican fishing practices and aspects of working-class opportunity.

Considered in this light, although Ernesto's familiarity with fishing developed while he was a child in Aguadilla, he established his fishing operation after he returned from New York at age thirty with money he earned outside of the community. Like Fernando, he did not begin his fishing operation from within, through a process known as endofamiliar accumulation (accumulation based on family and community participation in the fishing enterprise); instead, he began it with a sudden infusion of capital. Ernesto did not begin to fish commercially from a position in which the support of a community of fishers, or even a network or interrelated households, was absolutely essential. His attachment to that community, vis-à-vis association membership, is tenuous and even reluctant; yet he continues to belong because he perceives the benefits of collective action. Both before and after fieldwork for this study, many of the fishing associations of the West Coast joined together to protest new laws that regulate fishing and access to coastal and marine locations (Valdés Pizzini 1990a; see also Chapter 7). Aguadilla's association was particularly active in these disputes, its leaders having acquired speaking skills and political abilities from labor union involvement on the U.S. mainland. To support the association, even grudgingly, is in Ernesto's best interest.

Ernesto's experience suggests that long-term residence in one's natal community is not sufficient, in and of itself, to assure that a fisher will join his or her community of fishers with gusto. This is clear from other examples of prolonged community residence, even when the fishing experience is uninterrupted during that time:

In the North Coast community of Cataño, in an urban municipality within the San Juan metropolitan area, lives a line fisher named Hernando. Hernando's lifelong fishing partnership with a boyhood friend influenced much of what he knows about fishing and how he negotiates the social context of the local fishing community. Cataño was one of the key areas where shantytowns cropped up in the 1950s and 1960s, primarily to receive immigrants from other areas, such as Luquillo. On the boat of his friend and *compadre* Don Esteban, recently deceased, Hernando learned to fish at the age of fifteen. He was already familiar with fishing, because his father and grandfather had both fished in the canals and rivers of Luquillo, east of Cataño, when they were not working in sugarcane. But after Hernando experimented with free diving and harpoon fishing between the sixth grade and the time he was fifteen, Esteban taught him ocean fishing.

Until Esteban's recent death, he and Hernando fished with lines while their nets soaked. Now Hernando has a twelve-foot *yola* with a small motor, which he purchased after selling his eighteen-foot vessel. During Hernando's time with Esteban, the two men fished with gill nets for part of the year and with trammel nets at other times during the year. They found line fishing most productive during the summer months and during the month of February. For Hernando, fishing has always been a part-time activity, although, he says, it is "in my blood. It comes naturally to me. One looks for other jobs because fishing is not a secure source [of income]."

Despite Hernando's love for fishing, like his father's and grandfather's before him, he worked in sugarcane in Luquillo from the time he left school, after the sixth grade, until he turned fifteen in 1956. At this time, he moved from Luquillo to Cataño and took work as a taxi driver, both fishing and driving on a daily basis. In 1959 he added construction work to his means of livelihood, working as a carpenter's helper; yet he continued to fish as often as possible. During this period of multiple livelihoods, he also began to help his uncle buy and sell hogs, but this sideline lasted little more than a year. In 1965, he quit driving the taxi, and three years later he landed a full-time job with the Ports Authority helping out with the buses and other vehicles. Still, he continued to fish up to twenty hours during the week and another twenty hours on weekends.

After Esteban died, Hernando began to fish with Esteban's son under the same arrangements he had had with his friend: In contrast with the customary share system in which the catch is divided into quarters or thirds, with the boat receiving one share, Hernando split the catch (or the money from the sale of the catch) equally with his partner. Hernando now works only part time with the Ports Authority, which allows him to spend more time fishing.

There are no children in Hernando's house, but his wife, who is the same age as he is, helps him to prepare the catch for market and to sell it out of their home. As this independent fish selling suggests, Hernando has belonged to neither the fishing association of Cataño nor the fishing association of Luquillo. His only contact with the association of Cataño was when he once asked for help in getting a loan to purchase his own boat. The association failed to provide the help he needed, and neither he nor Esteban nor Esteban's son ever felt obliged to join.

It seems that, because of Hernando's long association with Esteban and his son, the small scale of Hernando's fishing enterprise, the work's apparent status as supplementary to Hernando's wagework, and the less-than-satisfactory experience Hernando had when he tried to secure a loan for his boat, he has not found it necessary to join his local fishing association. Moreover, in our conversation with him, we perceived few social linkages with fishers of the community other than Esteban, despite over thirty years of residence there. Throughout Hernando's life, his fishing has remained fairly simple: Although he has experimented with other types of gear, he keeps returning to the nets and lines, with the same crew, the same share system, the same small crafts. Hernando's case is also an example of how fishers look for jobs that have some of the characteristics of fishing—mobility, independence, and luck or chance—as well as jobs in the public sector, long believed to lend themselves to the retention of one's primary pursuit.

As is evident in the experiences of individuals presented thus far, becoming someone who can rely on fishing as his or her primary source of household income is never accomplished according to simple formula. Mere prolonged residence in the community, even when one continues to fish while there, is not sufficient. Association membership, without a broader nest of effective social ties in the fishing community, seems insufficient as well.

Equally interesting are these issues examined not as the influence of a person's social ties on becoming exclusively or primarily a fisher but as the influence of a person's combination of fishing with other occupations on his or her social links to the fishing community. This is, by extension, the broader issue of semiproletarianization's influence over the social and cultural context of fishing in general. Is wagework undermining the extent to which fishing, as an alternative to wage labor, is capable of social and cultural reproduction?

In Hernando's experience, fishers who control the Cataño association may be partially responsible for truncating some community members' social ties to the fishing community, actively limiting membership as a way of limiting pressures on marine resources and keeping state benefits (such as facilities and loan guarantees) to themselves. Such behaviors inhibit the growth of the fishing community from within. Hernando's experience may suggest that the practice of working outside of fishing allows association leaders to justify their refusal of assistance to those community members who fail to participate fully or actively in the association.

This interpretation comes from our general knowledge of Puerto Rican fishing associations, which often play pivotal roles in either cementing ties among fishers or fragmenting communities through perceptions and accusations of favoritism, privilege, and access to state support of fisheries. In the past, state support of fisheries has been critical during times of high unemployment and the decline of key coastal industries, particularly sugar; today state support is important for other reasons. In particular, state support provides legal backing in disputes with other government agencies that manage access to the coast or in disputes with private interests that privatize the coastline for their own uses. These private interests are becoming increasingly common throughout the Caribbean and other areas where tourism is on the rise, as fishers begin to compete with hotel chains for stretches of coastline or fight with park services and environmental interests that are scrambling to preserve mangroves and wetlands as resort communities bulldoze through nursery areas to fill in the salt marshes that percolate odors that are offensive to their guests.

In Pozuelo, on the Southeast Coast, we encountered the joint process of fragmentation and consolidation within the fishing community stemming from the fissioning of one association into two. The circumstances that led up to the break were typical: They began with disagreements over the marketing of fish and attempts by the CODREMAR-sanctioned association to control the association's facilities and the marketing behaviors of its members. Several of the fishers of Pozuelo broke from the association as a group, in a state of heightened class consciousness, and vowed to establish their own *independent* association with their own rules of fish vending and access to marine resources.

Yet leaving the association and its facilities meant finding an alternative location from which to operate, a location that offered facilities comparable to the association's or, at the very least, direct access to the sea. Unfortunately, the break occurred during a time of increased regulation of access to a coastline that was already becoming a playground for the rich. Painfully cognizant of this fact, the dissenting members found a polluted spot where, in years past, someone had dumped old machinery and where there was consequently no need to cut away protected mangroves. The members we interviewed fully believed that they were improving the coastal environment by clearing away obsolete machinery. They also believed that their work justified locating the new association on the cleared land. Clearly, they were squatting—they held no legal ownership to the land—but this means of property acquisition is not uncommon in Puerto Rico; in fact, it is sometimes even officially sanctioned by the local government through the provision of utilities and other public services. Generally, unless someone disputes the occupation, squatters remain and often even embellish their plots of land with elaborate, permanent dwellings. The new association members, who called themselves Pescadores Independientes, Inc. (Independent Fishers, Inc.), pursued the prospect of legal ownership: The association's president, Carlos Figueroa Villanueva, visited regularly with lawyers and local officials. Still, in the absence of legal ownership, they built the typical fishing association facilities: a pavilion with tables for cleaning fish, a ramp for launching boats, and a surrounding fence and gate with an elaborate sign bearing the name of the association. They printed up a leaflet with a statement of purpose and rules, which declared that the association had been established "to defend the rights and interests of the productive classes that live in rural areas and constitute the classes of fishers in Puerto Rico."

By this time, the old association, the one that had been established by CODREMAR, was nearly defunct. The president had quit, and several members—agreeing that the old association exercised too much control over its members' fish marketing practices—had followed the dissenters to Pescadores Independientes, Inc. One of the chief complaints, common throughout the island—was that the association insisted that its members sell their catch only through the association. The facilities were taken over by the current president, a stout, good-humored fellow who lets people know within the first few minutes of conversation that he and his family support the NPP. He avails himself of the facilities for his own personal use, running a restaurant out of its central building, and occupies the site with a handful of old members. He refers to Pescadores Independientes in highly derogatory terms. Three of his sons fish; yet his three daughters are pursuing careers in technical trades such as nursing, which suggests a measure of privilege that sets him apart from what members of Pescadores Independientes call the rural classes.

Two other circumstances make this story particularly telling: First, Ruperto, a prominent Pozuelo fisher (introduced in Chapter 3 and discussed in a related context in Chapter 5), belonged to Pescadores Independientes for a short time but quit in the first weeks of its existence. Second, Figueroa Villanueva's twin brother, Manuel, a highly successful trap fisher from a neighboring community, disparages not only his brother's association but also fishing associations in general. Not incidentally, both Ruperto and Manuel own fairly substantial fish markets. The former's is described in Chapter 3, and the latter's consists of large freezers in his garage, which sits above his fisher's workshop in a neighborhood of nicely furnished, middle-class, concrete homes.

Manuel's experience bears special consideration. Not only is his fishing operation successful; its success, like Ernesto's, was made possible by funds generated outside the fishing community. Unlike Ernesto, Manuel has extended his original capital with funds that, even if less capable of indebting fishers to him than some of the fish dealers of Puerto Real, seem very much like merchant capital.

At the heart of Manuel's fishing operation are over 250 traps that he and his men are nearly always in the process of building or maintaining in the open-air workshop beneath the garage and seafood market that adjoin his house. In addition to the traps, he uses a gill net, a type of gear he adopted in 1983 after an incident in which all his traps were stolen. Almost a full-time fisher now, taking only part-time construction *chiripas* when the fishing slacks off, Manuel nevertheless comes from a long history of wagework, having moved between seasonal construction work in Puerto Rico, seasonal farm work in Ohio, and factory work in New York. Of these, the construction work in Puerto Rico provided him with the steadiest employment. His job in New York lasted only two years, from 1967 (when he turned twenty-one) through 1968, and he picked sweet corn and cabbage in Ohio during the summers of 1972 and 1973, only two years after the birth of his first son. When his son turned eleven (around the time that most fishers begin to fish in earnest), Manuel quit working full time for the construction company and turned to fishing.

Like Ernesto, Manuel's capital for his fishing operation originated not from ties to other fishing households but from his work in the formal economy. Like Ernesto, too, at one time Manuel was a reluctant member of the fishing association in Pozuelo, the association from which his own brother split to head Pescadores Independientes. He and his twin brother quit the association at the same time, but Manuel did not join the new association.

Manuel's reluctant association membership is most likely related to the persistent disagreement between associations and fishers who have invested heavily in fish marketing, although Manuel adds to that a general bias against associations as forms of labor unions. During our interviews with him, Manuel spoke of his brother in somewhat disparaging terms and lambasted associations as groups of fishers who spend most of their time partying and complaining. Manuel also complained about the men he paid to fish with him. He referred to them more as hired hands than as fishing partners and claimed that he had to compete with government assistance programs in the hiring of crew, and still they demanded higher wages.

Manuel's comments on this subject, of course, derive from practical capitalist ideology—or, more bluntly, what we might call street capitalist ideology—that is, the standard employer's complaint against employees who fail to submit fully to capitalist discipline under employers' terms. In several studies of different branches of the food-processing industry and agricultural labor markets in the southeastern United States, in employers' personnel offices (from huge agroindustrial complexes in Georgia and Florida to small concrete-block crab houses hidden among the shell mounds of Tidewater, North Carolina), Griffith (1993) listened to the cries of street capitalist ideologists—ideologists informed less by theories of accumulation and political power than by daily business affairs. Over the past couple of decades, this street ideology seems to have percolated up into the halls of Congress—evidently a variant of trickle-down theory—culminating in the sweeping welfare reforms of the late 1990s.

For Manuel to subscribe to these ideologies is understandable in light of his position as a fish vendor. From the brief exposition of Puerto Real's fishery in Chapter 3, combined with other studies that show the importance of merchant capital in the organization and domination of fishing practices (Sider 1986; Vickers 1994), it seems clear that the migration of successful fish vendors toward such ideas reinforces political decisions that are designed to increase labor supplies and keep wages low. The new association in Pozuelo is designed to have a somewhat different effect: It creates social ties within a physical infrastructure that further differentiates the local economy, socially and politically, providing a context in which it is more common for fishers to respond to questions about crew recruitment and assembly with statements such as "No problem. I just use people from the community." Further dimensions of this opportunity are illustrated by the experiences of yet another member of Pescadores Independientes:

Humberto Hernández Ibarry, born in 1928, split from the
CODREMAR association with the first of the Pescadores
Independientes, working alongside those who invested their
labor in the clearing of rusty machinery from the site of the
new association. His work with the association came on the
heels of a spotty career in Puerto Rico and the U.S. mainland
working at several jobs. Like so many Puerto Rican men of
the era, Humberto began to work in the sugarcane fields in
1950, at age twenty-two, after several years of fishing with
his parents, both full-time fishers. After four seasons in the
cane, fishing during dead time, he was sufficiently indoctri-
nated into the rigors of workplace discipline to begin to
migrate to Pennsylvania to pick tomatoes. This experience
dampened his enthusiasm for wagework and, at the age of
twenty-seven, he quit work in the formal economy to fish,
like his parents, full time. He did not return to wagework for
twenty years, at which time—during the same year as the
birth of his daughter—he accepted a construction job in
Puerto Rico. Later he visited his son in California and took
a job as a janitor, first for Sears and then for a prestigious
federal agency. He worked at these jobs from ages 50 to 58,
leaving the agency after an injury claim that he settled for
three thousand dollars.

Humberto, who always relies on several types of gear—
including traps, gill nets, deep-water long lines, and trammel
nets that he takes joy in weaving himself—shifts between
fishing from his own vessel, a twenty-two-foot *yola* that he
acquired late in life, and fishing for shares on someone else's
vessel. As a *proel*, he receives 25 percent of the catch; as a
boat owner he receives 50 percent. Although he belongs to
the new association, he sells his shares of the catch out of
his house. He notes that it is possible, within the rules of the
new association, to sell one's own fish and still empower
one's class position with government assistance: *"Puede con-
tinuar pescando y vendiendo el producto de forma individual, y, para poder, recibir ayudas gubermentales"* ("One can
continue fishing while selling fish oneself, and, to be able to
receive government aid").

Humberto's political activism predates the formation of the association. He boasts of meeting Jesse Jackson in San Juan and speaks at length about the rights of fishers. His claim against the agency might also be considered an exercise of his basic rights of compensation for injury.

Humberto remains tied to the formal labor market through his wife, a woman eighteen years his junior, who worked as a nurse while they lived in California and who now works as a restaurant cook and helps prepare her husband's fish for market. Humberto's wife has nearly always worked in the formal economy, but the timing of her husband's forays into wagework seem as much related to household reproductive developments as to economic concerns. His return to wage labor after a twenty-year hiatus coincided with the birth of his daughter, and his subsequent work in California seemed designed to keep him close to his son.

By calling attention to both subtle and blatant differences between Humberto and Manuel, we return to points made earlier about membership, whether formally recognized or not, in fisher communities and the ways in which each person organizes his or her own fishing operation. Most notably, of course, the two fishers part ways in their association with and attitudes toward Pescadores Independientes: Each takes a historically opposing class position. More subtly, Manuel never assumes the *proel* or partner position on his vessels; he is always clearly the captain. In contrast, Humberto moves easily and without friction between the use of his own vessel and participation as a crew member on another's vessel. His relationships with other members of the community with whom he fishes are more similar to exchange labor relationships, or perhaps even gift exchange relationships, than to relationships between employer and employee.

Despite clear differences, both men agree that the marketing of fish is something best left to individual choice. Even so, it is likely that the two take this view from opposite class positions: Manuel wishes to use merchant capital to bind fishers to his own process of accumulation; Humberto fears being bound. Other members of Pescadores Independientes clearly share Humberto's concerns. Indeed, one of the principal reasons for the fishers' break from the old association was the organization's insistence on marketing its members' catches. Fish marketing in Puerto Rico, like food marketing everywhere (including by the establishment of restaurants), remains a central small-business opportunity among individuals with limited capital. Other fishers in Pozuelo who joined neither the old or the new association sell under a variety of arrangements—for example, Ruperto, the fish market owner who belonged to Pescadores Independientes for only a short while, who quit, most likely, after he realized that he could not become the principal market for the association. One of the oldest Pozuelo fishers we interviewed, a seventy-four-year-old net fisher named Oscar Garzán, has a long-standing arrangement in which he sells all of his lobster to a restaurant. But he sells the rest of his catch to his son, who fishes with him, and his son in turn sells this catch to dealers of his own choice.

While flexibility of fish marketing is, or at least is perceived to be, beneficial to fishers, the domination of marketing can benefit an association by adding to its facilities; improving its membership base; and garnering more political power, an advantage that most fishers perceive as central to association membership. Skillful association presidents need to balance these competing interests and resist the favoritism and self-interest that have plagued CODREMAR-established associations and that are associated with many government-assistance circles in Puerto Rico. Most (about two-thirds) of the fishers we interviewed did not belong to a fishing association at the time of our research; yet most (again, about two-thirds) had belonged to an association at one time in their lives. In places other than Pozuelo, too, we encountered cases of one association that fissioned into two. Thus, associations are commonly in a state of flux, either fostering or frustrating class consciousness.

Despite the precarious nature of associations within Puerto Rican fisheries, the fishers' life histories indicate that one barrier against waged work's subverting domestic production, as opposed to supplementing or complementing it, is a strong, healthy association. Whether or not it is state-sanctioned, a healthy association serves as a social infrastructure, a cultural setting and physical location where fishers can situate and elaborate other social ties in fishing communities (Robben 1989). Again, association membership, by itself, cannot enable fishing households to free themselves from wagework. Yet membership can aid in that process and in the process of reproducing households that are less dependent on wagework than were those before them. And reproducing households and communities is a process in which fishers must engage if they are to continue to provide a refuge for the fed up; the burned out; the retired; and (as Chapter 5 demonstrates) the worn out, who must dodge the everyday hazards of work at the beginning of the twenty-first century.

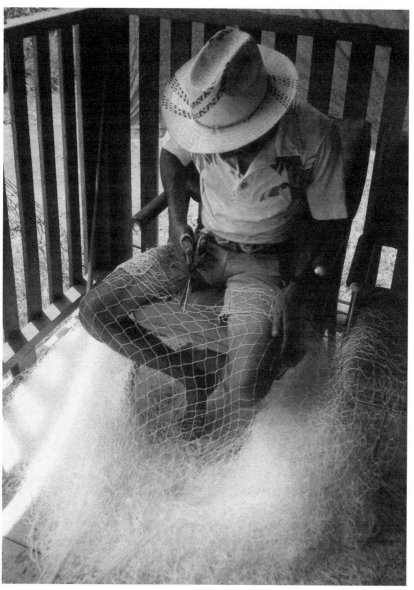

Puerto Rican fisherman splitting a gill net.

Fisherman entering calm waters.

Fishing sloop from Puerto Real (West Coast).

West Coast fisherman
with dolphin, barracuda,
and red snappers.

Snapper fisherman from
Puerto Real.

Scenes from the Mona Passage Snapper-Grouper Fishery.

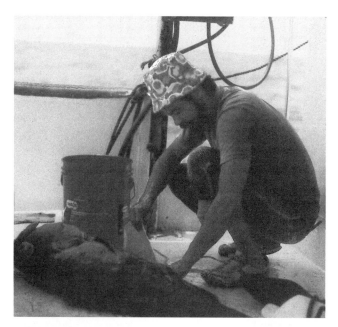

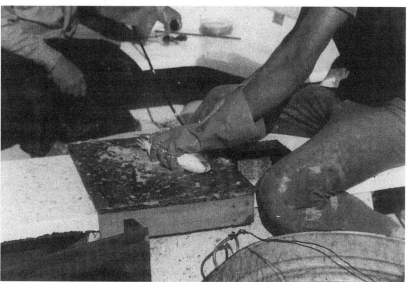

Processing fish at sea, Mona Passage.

Mending a cast net in Fajardo, a *villa pesquera*.

*Ceti* fishers in Arecibo (North Coast).

Fish traps (*below*) and lobster traps (*above*).

Shoreline in Puerto Real.

Combate shoreline, Cabo Rojo.

Typical *villa pesquera,* including lockers and a seafood market.

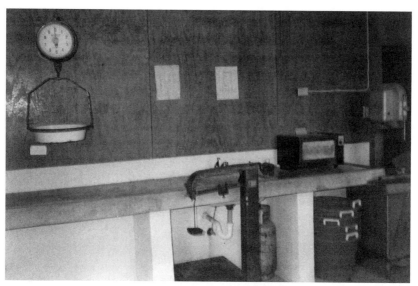

Interior of Culebra, a *villa pesquera.*

Puerto Rican fishing vessels in the shade.

Fishing families attending the Virgen del Carmen in Puerto Real.

# 5    Injury and Therapy

When I worked nights
on the milling machines
at Cadillac transmission,
another kid just up
from West Virginia asked me
what was we making,
and I answered, I'm making
2.25 an hour,
don't know what you're making, and he had
to correct me, gently, what was
we making out of
this here metal, and I didn't know.

—Philip Levine, *A Walk with Tom Jefferson*

ALONG WITH MATERIAL AND SOCIAL consequences of semi-proletarianization, discussed in Chapter 4, Puerto Rican fishers develop new and revised old conceptual categories to characterize their custom of moving among multiple livelihoods. These categories influence the ways fishers think about work as they influence the ways domestic production and capitalist production complement or contradict each other. Like many people around the world, Puerto Rican fishers have formed opinions about capitalism primarily from their experiences with low-wage, unskilled, or semi-skilled work settings (Frobel, Heinrichs, and Kreye 1980; Nash and Fernández-Kelley 1983; Sanderson 1985). Yet these opinions have not developed in a vacuum; they have been influenced by the prevailing ideologies of capitalism themselves and the power relationships that favor some ideologies over others—in other words, the processes of hegemony and counterhegemony (Gramsci 1971; Williams 1977; Wolf 1999).

Low-wage workers' attitudes about work and rewards draw heavily (and, at times, critically) on perspectives developed in ruling intellectual circles and dispensed through public schools, television, radio, and other media, along with more informal methods of communication. Rarely are ruling perspectives mere justifications for exploitation. Instead, they tend to be subtle and complex institutional and intellectual means of addressing how workers should live and experience capitalism (Rouse 1992). Further, they are rarely learned without being revised or reconsidered in light of workers' daily and lifelong circumstances. Within this process of revision lies the genesis of counterhegemony; or the struggle for a new hegemony; or a new, comprehensive method of dominating others' daily lives. As Williams suggests, "A lived hegemony is always a process.... It does not just passively exist as a form of dominance. It has continually to be renewed, recreated, defended, and modified. It is also continually resisted, limited, altered, challenged by pressures not all its own" (1977:112). Hermann Rebel clarifies this position by pointing out that the construction of a counterhegemony is heightened in times of crisis and duress, as during famine or the threat of starvation, injury, or death. Yet Rebel cautions against focusing on obvious crises to the exclusion of "the permeation of everyday life by culturally necessary ... victimizations" (1989:364).

In the lives of low-wage workers in advanced capitalist industries, these culturally necessary victimizations that permeate everyday life include a wide range of humiliating tasks and occupational injuries that we encounter in the life histories of Puerto Rican workers. Under advanced capitalism, many jobs held by labor migrants and factory workers tend to be hazardous, capable of injuring, poisoning, or killing workers. Both the International Labor Organization (ILO) and the U.S. Department of Labor rank jobs commonly held by Puerto Rican fishers as among the world's most dangerous. Although less-hazardous occupations have injury and illness rates of between 4 and 6 incidents per 100 workers, many low-wage jobs have rates that are two to three times higher: Construction work's injury and illness rate is 14.5 incidents per 100 workers, agricultural production's rate is 12 per 100, durable goods manufacturing's rate is 11.8 per 100, and food processing's rate is 16.2 per 100 (ILO 1983; U.S. Department of Labor 1987). Furthermore, these are only *reported* rates of injury. Occupational Safety and Health Administration (OSHA) investigations often find that underreporting of occupational injury is particularly widespread in hazardous industries (see, for example, North Carolina Division of Occupational Safety and Health 1989). For example, in recent work on meatpacking and poultry processing—industries that are notorious for employing minorities and immigrants—researchers have found that repetitive motion disease, cuts, and falls caused by slippery floors are principal worker concerns, points of contention in labor disputes, and issues around which workers organize for protest and legal action against their employers (Center for Women's Economic Alternatives 1989; Griffith 1993; Hage and Klauda 1989; Stull, Broadway, and Griffith 1995).

Responding to adverse working conditions, occupational therapy and vocational rehabilitation have developed and diffused programs and ideas throughout worker populations, providing new concepts and meanings that workers draw upon to characterize their worlds of work. Similarly, in other contexts, Western medicine has been used to explain and treat responses to the hazards of modern industry in third world manufacturing centers, usurping, complementing, or confusing native means of explaining and dealing with new industrial settings (see, for example, Ong 1988). Occupational injury is often aggravated by hazardous living conditions that are associated with low-paying work, particularly among migrants. Housing, for example, tends to be overcrowded, substandard, and subject to fire and vermin infestation, and transportation often involves unsafe, uninsured vehicles (Friedland and Nelkin 1971; Griffith et al. 1995). Many low-wage-worker populations occupy the same neighborhoods as groups engaged in drug trafficking, prostitution, petty theft, burglary, and life-threatening criminal activities (Griffith 1995). Furthermore, low-wage-worker households suffer the effects of inequities that are maintained and enhanced by legal mechanisms and cultural differences, including usurious interest rates; unstable or controlled (price-fixed) markets; differential citizenship status; discrimination in labor markets and courts of law; and, in the case of Puerto Rico, public school policies against developing English-language skills (Algren de Gutiérrez 1987).

Just as high rates of occupational injury have led to the growth of occupational therapy and vocational rehabilitation programs, so have the hazardous living conditions, discrimination, and inequality facing low-wage workers given rise to public and private social service and outreach programs. These programs dispense not only emergency funds, shelter, food stamps, medicines and medical services, legal services, and other forms of material support; they also provide new defenses and information with respect to the legal and political mechanisms that can remedy unacceptable living conditions, raise consciousness regarding workers' rights, and direct workers to labor unions or other organizations engaged in political struggle. Many problems of the poor receive additional attention in the context of liberal legislation, which justifies and offers legal support for civil disobedience, labor organization, and political activity that challenges existing social relations.

Responding to abuses toward farmworkers on the U.S. mainland, for example, the Puerto Rican Departamento del Trabajo (DT), or Department of Labor, pressed for the successful passage of Public Law 87, which guarantees through explicit legal contracts predesignated wage rates, minimal seasonal incomes, travel reimbursements, housing arrangements, and social security payments for all farmworkers that the DT refers to U.S. growers. In addition, the DT provides farmworkers with small blue books that describe, in both Spanish and English, their rights and their employers' obligations. Through experiences with programs and organizations such as these, small-scale producers have been able to glean ideas from the rhetoric and ideology of class struggle and apply them, sometimes haphazardly, to their own experiences. This occurs most swiftly under conditions of crisis, such as during strikes or other labor disputes (Edwards 1979; Hage and Klauda 1989; Mintz 1960), plant closings (Newman 1988), illegitimate developments in the quality of on-the-job supervision (Lamphere 1987), and severe cost of living problems associated with job loss (Nash 1985, 1994).

Although the incorporation of Puerto Rican fishers and other small-scale producers into wage-labor markets provides them with access to experiences and new ideas, the life histories show that this incorporation is uneven and incomplete. Most Puerto Rican fishers remain attached to natal communities and family fishing operations. As in many transnational communities, even after prolonged absence from the homeland and years of experience working abroad, Puerto Rican fishers participate in small-scale production systems and in the politics and economics of fishing in Puerto Rican coastal communities. This suggests that experiences in wage-labor contexts are received, understood, prioritized, critiqued, and acted upon through the cultural lens of Puerto Rican fishing.

## CONCEPTUAL AND POLITICAL CONSEQUENCES OF PROLETARIANIZATION IN PUERTO RICO'S ARTISANAL FISHERIES

### Semiproletarianization, Occupational Injury, and the Appropriation of Terapia

One conceptual response to semiproletarianization has been that fishers have appropriated the official rhetoric of occupational therapy and vocational rehabilitation, in particular the notion of *terapia* (therapy), revising its meaning in light of their concrete circumstances. In a survey of Puerto Rican commercial fishermen (Gutiérrez Sánchez 1982), two questions that illustrated fishers' sense of fishing as therapy elicited information about their appreciation of fishing as a job. One general, open-ended question asked respondents about why they fished, allowing them to express a wide array of reasons. The second question, translated from the Spanish, read, "Would you recommend fishing, as a job, to a young person you like?" (Gutiérrez Sánchez 1982:12).

In response to both questions, fishers launched into lengthy answers that used the word "therapy" and related their characterizations of the fishing experience. Revisiting the study, we evaluated the responses to the two questions already mentioned. Combining the responses, we found that 5.82 percent of 292 fishers actually used the word "therapy" to describe fishing; they made statements such as the following:

I am handicapped. . . . I receive social security [benefits]. . . . My family used to fish. . . . I earn some money. . . . It is a therapy.

[Fishing] has its advantages. It is a job as well as a therapy. One does not work for a boss, under the *servidumbre* (servitude) of no one, under no yoke. Nobody intervenes. It is only you and nature. There are people who work eight hours [a day] and earn a miserable salary; [in fishing] one works less. (Gutiérrez Sánchez 1982:14–17)

From the Gutiérrez Sánchez study, is it appropriate to say that fishing as therapy is a widely held cultural category among the Puerto Rican fishers, given the small percentage who actually used the word "therapy"? Respondents who spoke of fishing as therapy explained that its status as a "healthy activity" and a sport were key elements of the therapy of fishing. Extending these explanations, we find that 13 percent of the sample considered fishing a sport, and 10 percent of the sample defined fishing as a healthy activity. Thus, in all, nearly one-third (29 percent) of the fishers characterized fishing as therapy, as healthy, or as a sport. We emphasize that it is not our intention here to inflate the percentages in order to create a larger group of cases to discuss; it is our intention to show the semiotic extensions of the concept of fishing as therapy. Those who considered fishing a sport, for example, also described the activity as follows: "[It] keeps your mind occupied in useful things"; "it is a distraction"; "[it] keeps you away from drugs"; "it is entertaining, fun[,] it does not bore you"; "it is a clean activity"; "it is a good exercise" (Gutiérrez Sánchez 1982:14–17). Those who described fishing as a healthy activity made statements such as these: "It is good and healthy"; "it is peaceful"; "[it] keeps your mind occupied, away from bad thoughts"; "[it] clears the mind"; "[it] keeps you young"; "one forgets problems and tensions"; "[it] keeps you away from vices"; "it is a distraction"; "[it] keeps young people away from delinquency" (Gutiérrez Sánchez 1982:14–17). Although less than one-third of the survey respondents expressed their feelings about fishing in these terms, based on detailed life histories presented throughout this book, our contention is that the notion of fishing as therapy is widely held by Puerto Rican fishers, but the exact meaning of therapy has been altered to conform to fishers' concrete circumstances. Before we examine some of these cases, we consider the notion of therapy as developed in the lived and historical processes of capitalism.

*Therapy as a "Puerto Rican" Cultural Category?*

Historical studies of Puerto Rican mental health and social services institutions are a window on state strategies concerning specific conditions of the population *as a labor force* (Hernández 1985). These institutions are pivotal to the management of the population's health in relation to labor allocation (Bonilla and Campos 1985). Mental health institutions and professionals in Puerto Rico date from 1821, under the administration of the Spanish Crown (Hernández 1985). In the twentieth century, under U.S. colonial rule, asylums, psychiatric hospitals, and mental health services became so well established that the psychotherapeutic trends of the United States started to operate in Puerto Rico. The insular economy's dependence on large money transfers and social services programs from the United States established the general parameters for the dissemination of therapy as a cultural category (Pratts 1987). Although therapy was part of the language of mental asylums, it remained confined to the small, alienated, "mentally ill" portion of the population. It was not until 1963, when mental health clinics were established, that the concept of therapy (as used by occupational therapists) came into use in Puerto Rican households. Vocational rehabilitation programs, which had been in place since 1936, were redefined in 1969, giving the concept's popularization an added boost. Puerto Rican psychologists agree that issues that affect wage laborers—including the mental and physical conditions of the labor force, the treatment of migrants, the retraining of injured workers, and the rehabilitation of the lumpenproletariat (drug addicts, criminals, and juvenile delinquents)—have been a priority of these programs (Gómez 1985; Gonzalez 1985; Irizarry 1985). During the 1960s, the incorporation of people into the labor force became a clear objective of various government programs, based on the idea that workplace discipline was a cornerstone of social order. According to Irizarry, labor is "a mechanism of social control. In the job setting, socialization is consolidated, in correspondence with society's hegemonic values and ideology" (1985:164 [translated by the authors]). Labor is thus seen as rehabilitating, valuable to society as a whole as well as to the individual.

The establishment of vocational rehabilitation programs and two-year university degrees in occupational therapy consolidated social work as a profession just prior to the recession and oil crisis of the early 1970s and the subsequent large-scale return migration, during a time when the island's already overcrowded labor market was forced to accommodate even more workers. At the same time, however, industrial development expanded job opportunities, matching people with jobs through "vocational" programs in high schools and social services and attempting to control labor supplies by intervening in the reproductive behaviors of proletariat households (Bonilla and Campos 1985).

Within this broader context, social services that fishers were able to utilize included low-interest loans for the establishment of fish houses, training materials (gear, boats), and social security benefits for the disabled (cash used to buy fishing equipment). Channeling the young, injured, unemployed, and disabled into the economy, these programs were administered throughout the island, involving individuals from all segments of Puerto Rican society and contributing to the popularization of the idea of therapy. Among coastal peoples, fishing became therapy against unemployment, helping to maintain the domestic unit and reducing stress during hard times. This was well in line with the goals of occupational therapy, which included attaining low rates of unemployment, promoting good mental health, and facilitating the learning of new productive skills in order to adapt to changing opportunity structures (Gonzalez 1985:201).

Yet occupational therapy's aim of incorporating fishers into the Puerto Rican economy often contradicts the ways that the concept of therapy has been appropriated by Puerto Rican fishers. Fishers suggest that fishing is a therapeutic *alternative* to incorporation into the Puerto Rican economy *as a wageworker*, since wagework generally involves risks, separation from family (for migrants), and other hardships. As the life histories herein attest, most Puerto Rican fishers have worked in highly hazardous occupations; some have suffered occupational injuries themselves, and others have seen relatives injured on the job. A few of our interviewees were forced to leave school to seek work because of injuries incurred by the principal wage earners in their households. In the field, we encountered cases of the use of fishing as therapy (1) to help fishers mentally accommodate leaving the proletariat (including class-based political activity); (2) as a means of recuperating from on-the-job injury while earning income; and (3) as a vehicle for redefining political affiliations from the wage-labor proletariat to the fishing proletariat, contrasting fishing as therapy with wagework as injury and thus strengthening allegiance with others in the same work conditions.

Most important, however, while fishers receive physical and mental therapeutic benefits from fishing, they can continue to "work" without formally "working:" In other words, they can earn money without entering the formal economy, isolating themselves from political activity or altering the nature of that political activity. The therapeutic dimension of fishing, as contrasted with wage labor, is all the more important because working in wage-labor contexts also affects fishers' political activity *within* fisheries.

## THERAPIES AND TRAJECTORIES OF PROLETARIANIZATION

This chapter relates the following three trajectories of proletarianization to the metaphors of injury and therapy that fishers use to deal with their incomplete incorporation and the contradictions it entails: deproletarianization, proletarianization, and semiproletarianization. Some households, that is, may be reducing their reliance on wage labor, expanding their fishing enterprises (deproletarianizing); others may be increasing their wage-labor activity and phasing back or reorganizing their fishing (proletarianizing); still others may be maintaining, from generation to generation, a mix of wage labor and fishing (semiproletarianizing). The theoretical basis for the first two trajectories can be traced to Marx, who argued that, under the influence of capital, peasants would become either petty bourgeois or rural proletariat (1967).

Marx was not alone. The idea of the evolution of peasants and other domestic producers into definable capitalist classes also appealed to modernization theorists, who view this process as progress toward increased political and economic participation among people formerly isolated. Instead of drawing on Marxist notions of class or the rhetoric of conflict, however, modernization theorists tend to view peasants as practicing primitive production techniques and needing primarily diffused innovations to become modern, market-oriented farmers (Dalton 1971; Rogers 1969; see Roseberry 1983:chap. 7, for a critique of this orientation).

Whereas the first two trajectories have deep theoretical ties, the third (semiproletarianization) derives from more recent, primarily anthropological work on peasants and rural proletarians around the world. In both concrete and abstract contexts, these works have documented the long-term viability of peasant production while demonstrating that peasants are capable of combining wage labor and domestic production over long time periods (Collins 1988; de Janvry 1983; Griffith 1986; Mintz 1977; Roseberry 1988; Wolf 1982). In some cases, the expansion of capital has participated in the *formation* of communities and regions of small-scale producers, as well as many supposedly traditional customs (Rebel 1989; Roseberry 1983; Sider 1986). Of all three trajectories, semiproletarianization seems the most likely to generate the contradictory outcomes that the injury/therapy distinction attempts to resolve. Indeed, as we discuss further in Chapter 8, semiproletarianization seems particularly well adapted to late twentieth century and early twenty-first century capitalism.

We conceive of these as trajectories, rather than as specific mixes at one point in time, for at least three reasons. First, the direction of a household's proletarian status comments on qualitative features of wage-work opportunities for domestic production as the household passes through its life cycle. For example, when a household increases its wage-labor activity, this suggests that its domestic producer activities are insufficient to satisfy household consumer needs, or that the wage/benefit/working condition packages offered by the formal economy are more attractive than domestic production, or some combination of the two. The directional aspect also may signal a crisis in one or the other setting; this may be something as individual as a problem with the mobilization of household labor at a specific point in time or something as general and international as the Great Depression. Second, because a great deal of small-scale production is organized around changing household compositions (Chayanov 1966), the life cycle of the household—a temporal variable—may influence the mix of wage labor and independent production. Third, trajectories draw our attention to processes of social formation and dissolution, including the process by which social groups become subordinated to the needs of capital or struggle to free themselves from its influence, a struggle in which the appropriation and revision of conceptual and political material occupies a central position.

The following pages explore the ways that deproletarianization, proletarianization, and semiproletarianization come about and endure, their implications for fishing, how they influence class formation, and their implications for cultural analysis. Our specific focus here is how they influence the political and conceptual consequences of incomplete incorporation, particularly the association with fishing as therapy for the injury of wagework. We offer the following cases as *examples* of the ways that the three trajectories can affect household task allocation, relations among households, political activity, and other features of Puerto Rican livelihood. Exemplary fishers were purposely chosen from our broader, randomly selected sample; their selection was based on our knowledge of Puerto Rican fishing, the island's rural proletariat, and theoretical considerations (Johnson 1992).

## Deproletarianization or Reproletarianization?
### Two Related Trap Fishing Households on the South Coast

Deproletarianization, or withdrawing from wage labor to engage in commercial fishing full time, is a phenomenon that we could easily confuse with retirement, without the household and its life cycle as our primary reference point. The fisher who works as a janitor in New York from age nineteen to age sixty-five and then retires to Puerto Rico to fish cannot be considered deproletarianized; he or she remains a part of the proletariat, perhaps even continuing to receive some income (a pension) from his or her employment. Here, instead, we present the heads of two households who withdrew from wage labor while they were still relatively young. Our discussion of these households reconsiders the notion that deproletarianization implies moving toward petty bourgeois affiliations.

### First Household: Ruperto Correa

Ruperto, whose story was presented in Chapters 3 and 4, was the Pozuelo fisher who joined the newly formed Pescadores Independientes for a short time, only to quit after he realized that he could not use the association to improve his position in fish marketing. Ruperto's emergence from a proletarian background to establish a trap and net fishing operation that provides income to two households and supplies fresh fish to his own *pescadería* (fish market) represents one variant of the classic Puerto Rican artisanal fishing success story. Injury was a cornerstone of his success: He was awarded a seven-thousand-dollar settlement after being injured on the job at a multinational chemical corporation factory in his home municipality. In 1975, when he was twenty-six years old, he used this windfall to buy seventy traps, a twenty-two-foot boat, and two outboard motors.

The injury ended a thirteen-year history as a wageworker that had begun in 1962, when Ruperto was thirteen, when he moved from Puerto Rico's South Coast to New York with his parents. From 1962 to 1964, he worked as a bookbinder; then he waited tables in a New York restaurant until 1969. From 1969 to 1971, he served in Vietnam, after which he returned to Puerto Rico. Back home, he worked for two months in a small factory and then for nearly four years at the chemical plant, until his injury and settlement in 1975.

Like most Puerto Rican fishers, Ruperto came from a fishing household, although it was a semiproletarianized fishing household that (with Ruperto's parents' move to New York and his own employment history) seemed to be on a proletarian trajectory. Still, Ruperto's father and brother were fishers and Ruperto had begun fishing at age twelve, just a year before he moved to New York. Although his injury award enabled him to withdraw fully from wage labor, he had begun to prepare to return to fishing when he migrated back to Puerto Rico from New York, four years before the injury. Ruperto's journey was facilitated by Puerto Rico's tax-incentive program of industrial development, which lured many major conglomerates to the island. This succeeded in displacing much of the sugarcane production in Ruperto's home municipality, shifting the area's economic base from agriculture to manufacturing. At the same time, many laborers who had migrated to core regions began to return to peripheral regions between 1965 and the two world economic crises of the 1970s, as job opportunities in the core areas shrank.

Equally important is the fact that the community to which Ruperto returned was and still is a fishing community: It sits on a spit of land between a sheltered lagoon and the sea, with four boat access points on the lagoon, two fishing associations, and a recreational fishing club and marina. Most households in the community fish, even if only part time, and one of the associations (Pescadores Independientes) is currently involved in a dispute with the DNR concerning mangrove conservation versus access to the sea.

Ruperto thus returned to the social and cultural space of artisanal fishing, which—although helpful—was insufficient to establish a viable trap fishery. For that, Ruperto needed material support within his household and in his relationships with others in the community. He had relatives among other households of fishers in the community, and in 1975, when he received his settlement, he was able to move into commercial fishing with ease, hiring *proeles* from within the community to help him fish. As in many fishing households, in Ruperto's household—which includes his wife and daughter—his wife helps him construct and maintain the equipment, clean and process the fish for sale, and market the fish.

Nine years after Ruperto's injury settlement, he brought his cousin's household into his operation.

For the purposes of our discussion, there are four important features of Ruperto's fishing enterprise: (1) He owns his own fish market; (2) he drew his cousin's household into his operation; (3) his primary form of gear is traps; and (4) although he was a member of both associations at one time, he now belongs to neither.

SECOND HOUSEHOLD: JUAN CORREA

Juan, Ruperto's cousin, owns no fishing equipment himself and works primarily as a *proel* on Ruperto's boat, helping him to haul traps and nets. Instead of working under a share arrangement, the most common system used in the island's artisanal fishery, Ruperto pays Juan a wage according the amount of fish they catch. That is, Juan works more or less on commission, receiving thirty cents a pound for all the fish that he and Ruperto catch together and seventy-five cents a pound for lobster. Juan also helps construct and maintain the gear. Another benefit that Juan receives from the partnership is that Ruperto lets him use his expansive yard and his sets of tools (including welding equipment) to repair cars.

With the exception of the Ruperto's injury award, Juan's and Ruperto's work histories are similar. Both combine work in the United States with work on the island, and both maintain an attachment to artisanal fishing out of boyhood love and family heritage. Juan's father and brothers were fishers (as was at least one of his uncles—Ruperto's father). Juan, who is five years younger than Ruperto, worked as a migrant farm worker during the summers of 1973 and 1974, as a construction worker in his home municipality for four months in 1977, and finally as a mechanic's helper in San Juan from 1982 to 1984. He joined his cousin's fishing operation in 1984.

During Juan's times of unemployment, his father, brothers, and other relatives helped support his household. Two other members of Juan's household, his wife and eldest daughter, contribute to the fishing enterprise by helping to clean the fish, a service whose value should not be underestimated in an operation that is tied so directly to the market for seafood. The processing of the catch, both in terms of cleaning fish or creating new products—for example, *empanadillas* (seafood pastries)—is a job commonly performed by women. Juan also has two other daughters.

Like Ruperto, Juan joined the new, renegade association, breaking from the old association and helping to wage the access war against the DNR. Unlike Ruperto, however, Juan remained an active member and eventually became its treasurer.

The experiences of members of these two overlapping fishing households reveal several attributes of relations between wage-work and fishing. First, Ruperto withdrew from wage labor abruptly, using an injury claim against a major corporation as his initial investment in his fishing enterprise. Although Ruperto's case may seem exceptional, we encountered among fishers other cases of on-the-job injury claims, as well as a few variations on this "windfall" theme, including lottery, cockfighting, and other gambling winnings. Some fishers reported that they had to fish secretly, because they were receiving disability payments for injuries that they had claimed had made them unable to work. In one incident, a former policeman had claimed an injury, was caught fishing, and was then forced to return to the police force. Although the actual incidence of these windfalls may explain the source of investment capital in only a small percentage of fishing households, the stories are telling because they suggest that people from fishing backgrounds seek ways to leave wage labor in order to fish full time. Like the injury claims of individuals that are described in medical anthropology and political economy studies (Ong 1988; Taussig 1980), these injury claims emerge out of more general cognitive contrasts between the social relations of capitalism and native systems of meaning and production. In Puerto Rico, this contrast is seen in the association that fishers make between wage labor and injury and between fishing and therapy, beauty, tranquillity, and other healing qualities.

Juan's and Ruperto's experiences also illustrate the domestic character of trap fishing in Puerto Rico: It is based, fundamentally, on kinship (including close and more distant affinal and consanguineal ties) and tends to be the fishing style of choice among older full-time fishers. Although it is among the most productive fishing styles in Puerto Rico, trap fishing has a number of drawbacks that preclude large-scale entry by part-time or socially isolated fishers. Apart from the time spent in trap construction and maintenance (as much as four hours per day), one of the principal problems with setting traps is that they and their catch are susceptible to theft. The fact that a fisher must be able to trust his or her crew to keep the traps' locations a secret not only tends to keep trap fishing in the family but also constrains expansion.

In terms of class analysis, Juan's and Ruperto's experiences further illustrate that Puerto Ricans from fishing backgrounds will often leave wage labor to return to fishing even if it means substituting one form of subordination for another. Juan's with-drawal from wage labor only facilitated his taking the role of *proel* in Ruperto's fishing operation. Nevertheless, he substi-tuted a class-based subordination for one in which culture and class both participated. This formed a basis for Juan's political and social action, which Williams refers to as "lived":

> Cultural work and activity are not now, in any ordinary sense, a super-structure: not only because of the depth and thoroughness at which cultural hegemony is lived, but because cultural tradition and practice are seen as much more than superstructural expressions—reflections, mediations, or typifications—of a formed social and economic struc-ture. On the contrary, they are among the basic processes of the for-mation itself and, further, related to a much wider area of reality than the abstractions of "social" and "economic" experience. (1977:111)

The important observation here is that, despite the fact that his cousin/boss quit the new association, Juan not only main-tained his membership in Pescadores Independientes but also assumed the position of treasurer. This suggests that he was not willing to fully divorce himself from a class-based form of action. Further evidence of this comes from the new fishing association's charter, whose statement of purpose (translated by the authors from the association's brochure) quite consciously places the asso-ciation's membership within a broader group of rural producers:

> to defend the rights and interests of the productive classes that live in the rural areas and constitute the classes of fishers of Puerto Rico . . . to work for the strengthening of the country's fishing resources and for the material and social needs of the entire Puerto Rican rural population.

More abstractly stated, Juan's behavior shows that deproletar-ianization is not a uniform process of shedding one's class affili-ations to join the petty bourgeoisie of Puerto Rico. *That* form of deproletarianization seems more Ruperto's course. As an owner of his own fish market, Ruperto has to maintain good relation-ships with households that supply him with fish. Much of his political activity can be traced to his attention to marketing fish. We noted previously that Ruperto and Juan's community has two associations, one of which is currently fighting with the DNR over access. This association was founded not only to address the access question but also to break from and circumvent the market dominance of the government-built association. Ruperto originally joined the renegade association, but he quit after a few months despite the fact that his cousin was (and still is) an officer of the association. His decision to join the renegade association was understandable because his fish market competed with the market controlled by the other association. His decision to quit the new association suggested that he was not able to manipulate the association as a market and that the association maintained its integrity rather than submit to market dominance by a single individual.

It is also noteworthy that Ruperto's cousin remained an officer of the new association, maintaining a "lived" (or cultural) class affiliation with other fishers and (more abstractly) with the rural proletariat in general. *His* deproletarianization is thus more of a reproletarianization, or an attempt at merging cultural experience and class as a basis for political action. All of these market-related tensions, finally, are driven by the commoditization of marine resources, which is yet another way that capital participates in the formation and disruption of community alliance.

### Proletarianization

Our sample selection procedures precluded us from studying households that did not fish. Nevertheless, we can describe the process of proletarianization—or the complete withdrawal from fishing to become part of the rural proletariat—that occurred between generations in some households. Most such households were from fishing communities that now have weak attachments to fishing and whose children do not fish at all. They are semi-proletariat households; yet they are not reproducing this status: Their future generations are likely to be fully proletariat. Those with weak attachments to fishing, moreover, often regard it as a subsistence or recreational enterprise or, more pertinent here, as a therapy.

THIRD HOUSEHOLD: HÉCTOR MUÑIZ

Héctor, who is in his mid-seventies, has had a complex labor and small-scale production history, of which fishing occupies a small but significant part. His main fishing gear consists of hook and line, although, in keeping with his generalized economic strategy, he also uses a few crudely constructed traps and nets. He fishes, usually alone, primarily for subsistence, selling little of his catch. Although there is a fishing association in Héctor's town, he is not a member, and the association is not very active in the community. He now lives with his wife; their five children have left home without any intention of expanding Héctor's fishing enterprise.

Héctor's labor history began in the sugarcane fields of Puerto Rico, an occupation from which he never fully escaped. Between seasons in the cane fields, he worked in construction in his home municipality, worked on *fincas* (farms) in New York and Florida, and washed dishes in a restaurant in Miami. After years in the states, he returned to the Puerto Rican cane fields, a decision guided by one of his sideline economic activities and the central love of his life: raising cocks for cockfighting. A tour of his work space—a garage with a ventilated roof attached to his house, along with the cages, cocks, and tools of his trade— gives evidence that raising gamecocks is his passion. He keeps cocks at all stages of development, but none are for sale. With their nipped combs, punctured heads and beaks, spurs wrapped with adhesive tape, and thinning feathers, they look as though they have fought plenty. Héctor keeps a locked box that contains a small leather case that holds spurs taken from dead birds; these he polishes and sharpens to look like small translucent yellow, red, and tan antlers. If a cock's spurs are not well developed, he fastens the dead cock spurs to the stubs. Despite all Héctor's enthusiasm, care, and attention to his gamecocks, according to his wife, the cocks do not generally do well; if they lose, he brings them home and she "makes soup out of them."

Héctor's wife also keeps in the garage the tools to tend a small vegetable garden just across the road from their house. She grows vegetables primarily for home consumption. Scattered among her tools and his gamecocks are the crude fish traps and other gear that he uses to fish.

Unlike the Correa cousins, Héctor did not return to a fishing community when he came back to Puerto Rico from his stay on the U.S. mainland. Instead, his community is a dense cluster of homes in the central town of a rural municipality. The municipality itself is surrounded by sugarcane fields. Héctor's barrio has house after house, one nearly on top of the other, each with a small concrete or dirt yard. One of his neighbors cooks tubers in a big pot and mashes them into a paste for sale on the street. From neighborhoods such as these, come not only street vendors and others who engage in the informal economy but also public works employees, sugarcane workers, and those who find employment by joining the migration streams along the U.S. eastern seaboard.

Héctor's household differs from deproletarianizing households in three important ways: First, when he returns to Puerto Rico after migrating north, he settles not in a fishing community but in a heavily proletarian, small-scale-producer community (a community whose economic base is informal economic activity) in which fishing is only one of many survival strategies. A second feature that differentiates proletarianized from deproletarianized households is, obviously, their marketing behavior. Hector sells very little of his catch and thus has none of the burdens of marketing that Juan and especially Ruperto have. And he does not expend any political energies for marketing either.

Finally, the space that Hector devotes to his fishing enterprise is smaller than what is typical among full-time fishers, and it is cluttered with the paraphernalia of a number of domestic producer operations, notably his gamecocks and his wife's gardening tools. This illustrates an important contrast between the allocation of work space among full-time and part-time fishers: Both the size and character of the work spaces tend to distinguish households that engage in full-time artisanal fishing from households that fish on a more casual, subsistence or recreational basis. In contrast to the amount of space taken up by fishing gear in Héctor's small garage, the largest portion of Ruperto's yard is devoted to gear construction and maintenance. Structurally, Ruperto's fishing work is more sound than his living space; the former is enclosed by concrete, and the latter is made of wood. Ruperto's fish market is also made of concrete, is nicely painted and decorated, and has bars over the openings. These fishing work spaces are central features of artisanal fishing on the island. Generally, they resemble garages at service stations in the United States, cluttered with fishing tools, lubricants, and equipment—oily, dirty, sour smelling, and solid. Héctor's garage, in contrast, resembles more a barn at a children's petting zoo. Its appearance reflects (1) the way that Héctor and his wife combine various subsistence activities and (2) how these activities fit into their otherwise proletarian existence. Though the gardening, cockfighting, and fishing provide subsistence and modest incomes, they also double as hobbies, sports, and recreational activities. In this regard, these activities act as therapeutic relief from hazardous, difficult, and tedious jobs without entangling Héctor in political thought and action as a small-scale producer. A look at other proletarianized households leads to similar conclusions:

### FOURTH HOUSEHOLD: CARMEN MARÍA NAVARRO

Fishing in Puerto Rico is primarily a male enterprise, and in fact Carmen's original attachment to fishing and to the fishing community derives from her husband's experience as a fisherman and from kin ties with male fishers in the community. Throughout Carmen's husband's life as a wageworker, he fished only on weekends and only for recreation; on occasion, Carmen would accompany him. Recently, however, Carmen's husband died, and she sold the fishing equipment because it brought back too many memories of her husband.

Although they lived in a fishing community and had kin ties to fisher households, Carmen, her husband, and all but one of her children (a nine-year-old son) spent the bulk of their economic lives in wage-labor jobs. They are, clearly, proletariat. All of her older sons and one of her daughters live and work in New York; another daughter lives in Boston, and two others live in Puerto Rico. All work in factories. Only one of her sons attempted to launch a career in fishing, but he eventually left fishing to engage in wagework in New York like his brothers. Carmen herself worked as a domestic servant in San Juan; she was recruited for this job as part of a government program to locate and place rural women into domestic service jobs in the metropolitan area.

Like Héctor's complex mix of household economic/subsistence/recreational activities, Carmen's fishing, wagework, disability and social security payments, and other economic activities are enhanced by other, informal economic activities, including coordinating festivities for the local fishing association and the social functions of other businesses. In addition, in the community, Carmen is known for her skill as a *bruja* (witch). She uses her craft to protect some neighbors and avenge others.

Despite complex economic and cultural activities such as these, during one point in their life together, Carmen's and her husband's fishing assumed a dimension that was far more important than simple recreational activity. A more serious approach to fishing emerged from an injury and therapy context. Carmen and her husband began to fish together as a therapy after her husband, a heavy equipment operator, suffered an on-the-job injury. Jumping from a huge vehicle, he injured his spine so badly that he could no longer engage in any heavy work. Carmen explained that, after the injury, her husband began to receive disability payments; yet they began a more regular fishing agenda *"como una terapia"* ("as therapy"). Their primary gear consisted of a gill net, a fourteen-foot boat, and a six-horsepower motor. These they used together full time only between her husband's injury and his death in 1988, after which Carmen gave up fishing entirely.

Here fishing satisfied rehabilitative functions, occupying Carmen's and her husband's time with an economic activity that did not require excessive strength and supplemented his disability payments. The fact that Carmen sold all their fishing equipment and gave up the activity after her husband's death reflects the fact that their involvement in fishing did not include joining a society of fishers (an enduring basis for political action, a class). This is especially interesting in light of Carmen's many kin ties among other fishing households and her active involvement in the fishing association as a caterer or organizer of social functions. Although she has ample opportunity to involve herself in the politics of artisanal fishing, she has not done so. Instead, during her husband's life, they appropriated the notion of therapy not as a means to locate themselves in a broader group united by common suffering from occupational injuries or joined in struggle but as a model for rehabilitation in the aftermath of injury. In addition to using fishing explicitly as a therapy to relieve injury, Carmen and her husband used the activity in much the same way that Hector did: as an escape from the hazards and difficulties of low-wage labor.

In its broadest features, proletarianization reveals little that we could not have easily predicted: For these households, the opportunity to make a living from fishing does not exist. Either they perceive the resource to be an inadequate material base, or their means of exploiting the resource are constrained by social or economic features of their lives. Their children usually become wageworkers, often outside the community, and they have not embraced the kinship or network ties between two or more households that underlie domestic operations such as large-scale trap fishing.

By drawing political and economic energies away from fishing communities, external opportunities undermine diffuse, meaningful community ties by allowing individuals to survive without them, fragmenting communities and production operations in the process. Local economic opportunities cease to become central to local ways of life. More concretely, here fishing gradually becomes less important in the household's overall survival strategies, taking its place alongside a number of wage-labor and domestic producer activities. Each contributes to total household income, but none either emerges as an identifying or central economic activity capable of sustaining the household or motivates political activity. Political energies are expended, instead, either on getting and keeping jobs in the public sector through political patronage or on more traditional working-class-based political activity, such as joining trade unions or soliciting the aid of state or nonprofit agencies that support the rights of labor.

*Semiproletarianization*

Most Puerto Rican fishers' households have the weak attachments to the proletariat that characterize semiproletarianization, either through their own experiences or through relationships with others who have direct, substantial ties to their households. With the exception of two or three fishing communities on the West Coast, most fishing communities in Puerto Rico are semiproletariat communities. Again, we emphasize that the semiproletariat condition is important not only because it is so common in Puerto Rico and around the world but also because, of all three trajectories, it is the one most likely to generate the contradictions that the injury/therapy distinction attempts to resolve.

### Fifth Household: Julio "Marty" Seba

A part-time net fisherman on the Southeast Coast, Marty fishes with his brother, who—like Marty—has a complex occupational history. They devote around one-third of their work time to fishing. Marty worked for one year as a ticket writer in the cane (the person who assigns each row of sugarcane a value, the amount to be paid to the cutter). Then he worked in the United States, first doing farmwork and later in a turkey-processing plant in New Jersey. After that, he worked in construction for three years, and from 1974 to 1986 he was employed by the local fishing association. Finally, he landed a position with the local government.

Marty's occupational history, like his brother's, revolves around local political developments and fishing. For example, because his clerk job for the municipality depended on political connections and his brother was working as a mason on construction jobs for the municipality, both anxiously awaited the outcome of the 1988 election. If their candidate lost, they intended to migrate once again to the U.S. mainland in search of work.

Marty and his brother both live with their mother; children from their failed marriages live with them. Marty has a seventeen-year-old daughter in high school, and his brother has an eight-year-old son. For the past sixteen years, Marty and his brother have used nets as their principal gear, but this varies throughout the year as different species migrate through the bay where they fish. Recently, Marty bought a 150-foot-long gill net, an expensive piece of gear that is often repaired by someone other than the fisher. Marty commented about his brother, *"Decidió pescar porque le gusta y porque considera esta actividad un 'hobby' o una terapia"* ("He decided to fish because he likes it and because he considers this activity a hobby or a therapy").

The busiest fishing season for Marty and his brother is between January and May. Rarely do they fish every day; often, they fish two to three days per week, on weekends, or in the afternoons after they finish work. As with many fishers who use the gill net, Marty receives half the catch because he owns the net and the boat that they use to set the net. The remainder is divided among the members of the crew—usually his brother, the young nephew who lives with them, and a friend. The division of the catch changes from time to time, because sometimes Marty's brother helps pay for net repair and maintenance.

Although Marty used to sell fish through the association, and in fact was one of the association's principal market people, he now sells his fish (usually at least 85 percent of the catch) in the street.

Like the deproletarianized fishers previously portrayed, Marty lives in a fishing community and has been active in the fishing association in the past (in fact, he was among the founding members and at one time held a paid position in the association). This community, however, has only one association, a strong organization that most of the community members belonged to until the year prior to our research. At that time, the association fell apart when a dispute between the president and CODREMAR caused a change in association leadership. Currently, the association is all but defunct; one family operates a restaurant on the top floor of the facilities.

One observation we can make here is that Marty's association membership, which included a paid position, preceded his employment with the municipality. Evidently, he benefited from political skills gained while he was with the fishing association and was able to obtain a job in the public sector through political ties. For some of the other fishers we interviewed, this sequence was reversed: Political skills learned through wage-labor experiences helped in the development of leadership skills that, in turn, aided the political struggles of fishing associations.

The interplay of political skill between fishing and the formal economy is all the more important when we consider that neither Marty nor his brother has ever been a full-time fisher. At the same time, neither intends to divorce himself fully from fishing. Furthermore, Marty's nephew (his brother's son) is learning artisanal fishing skills, becoming prepared to use fishing either as a supplement to other incomes or, perhaps, as his principal income producer. These preparations may serve to reproduce the household's semiproletariat condition.

Marty's own characterization of fishing as a hobby or therapy makes his experience all the more complex, suggesting that fishing can relieve the tensions of wagework as it simultaneously serves as a basis for political activity. Here again is an example, like Juan's household, of a case in which fishing assumes a complex, "lived" role that draws upon both culture and class. And here again, Williams's observations, following from the passage cited previously, are relevant:

> People seeing themselves and each other in directly personal relationships; people seeing the natural world and themselves in it; people using their physical and material resources for what one kind of society specializes to "leisure" and "entertainment" and "art": all these active experiences and practices, which make up so much of the reality of a culture and its cultural production can be seen as they are, without reduction to other categories of content, and without the characteristics straining to fit them (directly as reflection, indirectly as mediation or typification or analogy) to other and determining manifest economic and political relationships. Yet they can still be seen as elements of a hegemony: an inclusive social and cultural formation which indeed to be effective has to extend to and include, indeed to form and be formed from, this whole area of lived experience. (1977:111)

Economic and political relationships do participate, quite fundamentally, in the way in which Marty receives fishing into his life. This is revealed in the most concrete features of his fishing operation, where semiproletarianization influences his choice of fishing gear and other fishing features. The seasonal and afternoon use of the gill net, the way that the bounty is divided, the fact that the catch is landed daily, are some of the characteristics that fit well with Marty's wagework. Yet Marty's fishing operation is also constrained by his wage employment and by his lack of a substantial kin network to allow expansion. Although his fishing depends on consanguineous ties (his brother and nephew), these ties have not resulted in the formation of the separate yet linked households that characterize the deproletarianized households previously discussed. In the sixth and final case that we describe, these separate yet linked households have been combined in a way that gives semiproletarianization another face.

SIXTH HOUSEHOLD: ESTEBAN REY

Esteban lives in a small, bare, concrete-block unit in a new, government-subsidized resettlement project, having been forced inland a few miles from the South Coast after flooding destroyed his previous home. He is thirty-six years old, married, and the father of three small children. His mother and father live near the coast, and one of his brothers works in a factory in New York.

Esteban has worked in the formal economy both regularly and sporadically. He has taken farming jobs (harvesting tomatoes) and has done related agriculturally based jobs, such as cannery work (which his mother also did for a number of years). But, with one exception—an eighteen-year part-time job at a pharmacy—his jobs did not last long. Esteban belongs to no fishing association, explaining that his pharmacy job kept him too busy to join.

Esteban's mother and his brother in New York control the bulk of Esteban's participation in artisanal fishing. Like Ruperto, Esteban's mother escaped from wage labor into artisanal fishing by means of a windfall. With fifty thousand dollars that she won in the island lottery, she bought coastal property and traps and established herself as the matriarch of the family fishing operation. Playing and winning the lottery conforms with Esteban's parents' other gambling/risk behaviors. His father shoots pool and plays cards to supplement their income, leaving the mother with principal control of the fishing enterprise.

The family has over one hundred fish traps, their principal gear. Esteban, the primary fisher, sets traps one day a week for his mother and one day for his brother, diving around the traps for lobster, conch, and octopus. Esteban's mother sends money from the sale of the fish to her son who lives in New York. For his part, Esteban receives part of the catch, has access to the family's fishing equipment, and has the opportunity to earn money from an activity that he obviously thoroughly enjoys.

It is evident that fishing is more than just a job to Esteban from the great animation with which he describes his diving experiences; he is fond of telling about the dangers of the deep, gesturing wildly as he recounts tales of wrestling octopus from reefs, escaping sharks, and spearing fish. In other accounts, he characterizes his experiences underwater in terms of the quiet and tranquillity.

Esteban's family sells nearly all of its catch at a small stand near the water. Usually the fish are sold whole, but sometimes Esteban cleans the catch, which earns him ten dollars per hundredweight of cleaned fish. Esteban's father does not usually fish, but he supervises the sale, since most of his gambling activities take place at night.

Esteban's story shows how fishing households can become semiproletarizanized by virtue of their relationships with other households. Three separate yet linked households are involved in the fishing operation: Esteban's, his parents', and his brother's (even though his brother lives in New York). It is telling that, while the brother in New York works full time in the formal economy, he makes production decisions that impact the other two households. This brother's position is structurally similar to that of his mother, particularly in relation to Esteban's household.

This account shows that the semiproletarian condition need not necessarily constrain decisions regarding gear, *as long as* the households are part of kin networks that include individuals who have the time to handle gear that requires additional time and effort. The integrity of the fishing enterprise remains, including its value as entertainment, its perceived rewards, and its ability to contribute to the maintenance of wageworkers' households.

This case also underlines the importance of gambling in Puerto Rican fishing households, demonstrating that shifting from one trajectory to another often requires the good luck of a windfall. As in Ruperto's case, prior to the injury settlement, the family seemed to be following a more proletarianizing path than its current course, which successfully combines full-time participation in the formal economy with artisanal fishing.

## INJURY AND ITS THERAPY

Puerto Rico occupies a somewhat unique position in the international division of labor, as well as in the history of North American anthropology. As Spanish-speaking U.S. citizens, as citizens without voting rights or full political representation in Congress, as a people culturally distinct from other U.S. minorities, and as objects of discrimination, Puerto Ricans remain second-class citizens—a people apart from yet a part of North American capitalism (Duany 2000; Morris 1995). Since the United States appropriated the island from Spain in 1898, and especially since the U.S. government granted Puerto Ricans access to the U.S. labor market in 1917, Puerto Rican society has served as a source of low-wage labor. Workers could be easily recruited to U.S. jobs and almost as easily "repatriated" during times of recession and depression (Bonilla and Campos 1981; Steward et al. 1956; Wessman 1977). This has meant that several Puerto Rican social groups, including artisanal fishers, have been unevenly incorporated into North American capitalism and have thus engaged in struggles for improved terms of trade between their households and capitalist markets.

In light of this condition, it is no coincidence that Puerto Rico hosted some of the first and most influential studies of capitalism's role in the growth of communities. We noted in Chapter 1 that one of the hallmarks of this work was the project headed by

Steward in the years following World War II, although we also pointed to problems with Steward's conception of Puerto Rican "levels of sociocultural organization" (Steward et al. 1956:6–7). Specifically, Steward's concept was designed to explain neither interaction among levels nor a radical departure from traditional/modern paradigms nor folk/urban continuums nor other typological work that viewed increased wage labor among small-scale producers as symptoms of a general transition from preindustrial to industrial society (see, for example, Dalton, 1971; Redfield 1947; Rogers 1969; Roseberry 1978:28; Tonnies 1955). Others who participated in *The People of Puerto Rico* project, however, particularly Mintz and Wolf (Steward et al. 1956), made some of the earliest contributions to the idea that capitalist expansion could participate in the formation, maintenance, and reproduction of so-called traditional social, cultural, and technological adaptations (Roseberry 1983; Sider 1986). According to Roseberry:

> Unlike later world-system theories, the goal of historical investigation was not to subsume local histories within global processes but to understand the formation of anthropological subjects ... at the *intersection* of local interactions and relationships and the larger processes of state and empire making. ... Even as their objects of inquiry changed ... the interest in the formation of anthropological subjects at the intersection of deeply rooted local and global histories remained. (1988:164)

The trajectories outlined in this chapter demonstrate the diverse responses of households, communities, and domestic production operations to capital's demand for low-cost labor. This diversity, moreover, is what one would expect if capitalist expansion resulted in intersections of "deeply rooted local and global histories."

Beyond this, however, the trajectories illustrate how variegated and ongoing the processes of proletarianization are. Interactions between capitalist and domestic processes are diffuse, varying settings and times yet continually drawing upon and revising material and ideological realms. These processes interweave, touching one another at very real, human points of interaction: migration, the commoditization of marine resources, and disputes over coastal access. They further influence one another through symbolic reflection, as when fishers appropriate concepts such as therapy from capitalist-based systems of meaning and invest them with new meanings.

The diversity of responses to capitalist domination, moreover, confuses the logical consequences of political consciousness (that is, class formation) by involving a variety of concrete manifestations—such as different gear types, different relationships within and among fishing households, different uses of social and cultural time and space, and different degrees of involvement in organizations (such as fishing associations)—that may serve political ends. Equally important, each trajectory involves different *uses* of artisanal fishing itself: as *primarily* an economic activity, *primarily* a therapy, or *primarily* a therapeutic supplement to the household's full range of (mixed) economic activities. A household's or individual's attempt to shift from one trajectory to another, diverting fishing to new ends, encounters difficulty precisely because some types of fishing gear work more effectively with specific social units and specific allocations of time and space than do others, and they require different levels of dexterity, experience, and technical knowledge. Furthermore, these "optimal" social units, time and space allocations, or necessary skills depend on an individual's or a household's participation in the formal economy and its place in relation to other households engaged in domestic producer operations.

This last point is crucial. A *community* of fishers—or at least a community of a network of separate yet linked households—is a necessary mechanism for shifting from one trajectory to another in a way that will improve the household's terms of trade. The separate yet linked households we have seen here may be understood as a creative means of dealing with a changing international division of labor that relies extensively on internal and international labor migrations. As we see in many studies of labor migration, domestic units typically conform to structures that are most useful to capitalism, compensating for shortcomings by modifying their articulation with the formal structures of advanced capitalism (Basch, Glick-Schiller, and Szanton Blanc 1994; Griffith et al. 1995; Mahler 1995; Massey et al. 1987; Portes and Bach 1985). We see this in Esteban's case in particular, in which his extended family spans three separate yet interdepend-ent residential units. The domestic unit has been able to (1) main-tain its integrity as an extended family by substituting benefits that derive from residence (for example, labor, skills, emotional and political support, and technical knowledge) with remittances to and from the mainland and (2) anchor its incorporation into capitalism in household-based, small-scale fishing in Puerto Rico.

In these households, class formation becomes compromised as the domestic units, social networks, associations, and other social forms serve critical social purposes such as reproduction, health care, protection, and social security that confuse the foundations upon which classes form and develop. These social forms also serve critical cultural purposes as vessels of values and myths that can either envelop advanced capitalism in its protective symbolic skin (for example, the notions of risk or the free market) or criticize its consequences as contrasted with domestic production (for example, injury versus therapy). Among Puerto Rican fishers, we encounter a pervasive desire to reduce their dependence on wage labor, a desire that leads some of them to portray wage labor in terms of injury and sickness and fishing in terms of therapy and healing.

# 6 Roads Less Traveled

## *Proletarianization and Its Discontents*

INDIVIDUAL CHOICES that seem to run counter to prevailing structural conditions tend to stimulate discussion about exactly how and how much those conditions influence opportunity and experience. In the many stories that constitute an individual's life history, specific choices often emerge as critical explanations for the routes that lives take vis-à-vis personal, social, and economic crises. Yet the choices that individuals make, while personal in essence, also respond to barriers that derive from customary economic and political practices, including the structure of the labor supply at an international, global level. This chapter illustrates how individual choices, made within household contexts, combine elements of the structural and the personal in a manner logically similar to Marx's notion of the dialectic between ideology and behavior.

## CRISES AND THE DISENGAGEMENT
## FROM SUGARCANE PRODUCTION

Let us look at Santos's experience:

> Born in 1944, Santos is a fisher who lives with his wife and five sons. He learned his trade at age seven, perhaps earlier than most Puerto Rican fishers, in what many fishers come close to explaining as a symbiotic process. Living close to the ocean, they had the opportunity to learn by watching others fish. Because no one in Santos's family fished, he learned the trade with friends from the community.
>
> A school drop-out at age ten, Santos became a sugarcane cutter in 1954, first in a sugar mill in Camuy, later operated by the government's Sugar Corporation, as well as for private farmers who grew sugarcane on contract for larger farms.

Beginning in 1961, Santos worked as a migrant worker on farms in New Jersey, New York, Pennsylvania, and Connecticut, performing diverse tasks such as land clearing, fumigation, and harvesting. Traveling between the U.S. mainland and Puerto Rico as a cyclical migrant, Santos worked simultaneously for the private farmers, as a migrant worker, and in the sugar mill until 1973, when the mill ceased operation.

Like Santos, most other fishers who worked in the cane left sugar production in the 1960s and 1970s but did not enter fishing immediately. They took a long route that included proletarianization in other sectors of the insular economy and in the United States.

Along with other fishers/sugarcane workers, Santos navigated the troubled waters of a socioeconomic process that affected most Puerto Ricans and certainly the insular economy as a whole. As we noted in Chapter 2, following the demise of the sugarcane industry after 1940, the Commonwealth of Puerto Rico embarked on a modernization plan, courting U.S. capital. Transfer payments, industrial development, improved health and fertility planning, and the state's bailout of failing sugar companies were among the initiatives; yet together they could not prevent the dislocating forces of unemployment that encouraged rural to urban and Puerto Rican to U.S. migration all across the commonwealth (History Task Force 1979:141).

Although embedded in these processes, Santos characterizes his choices as motivated by personal experience. As we discovered in many of the life history interviews, Santos was drawn, initially, to the therapeutic qualities of fishing. His deproletarianization, stimulated by the workings of the sugarcane industry and its demise, was accelerated by a psychological or neurological condition. As Santos stated (and as was observed in the field), he is ill and is *"afectado de los nervios"* (affected with a nervous/neurological condition), which explains his ultimate disengagement from wage labor.

At this time, the demise of sugarcane production was underway yet incomplete, and we can establish no landmark date that marked the death of its central position in Puerto Rico's economy. Having pointed out that the 1960s and 1970s represented decades of major decline in production, as well as commonwealth support through the Corporación Azucarera (see Giusti Cordero 1994: 1008), let us visit a previous decade, when different circumstances came to life on the island of Vieques, east of Puerto Rico. In Vieques, his voice and that of Victor add to Santos's case an illustration of a critical period in which the interests of proletarian households clashed with the objectives of sugarcane companies and the hemispheric defense planning and development of the U.S. Armed Forces.

In 1922, when Santos was eight years old, he and his brother migrated from the eastern coastal town of Ceiba to the island of Vieques in search of work. Vieques then had a rich fishery, and employment opportunities in the sugarcane industry were abundant. Both brothers started to work in the sugarcane fields coincident with their arrival, child labor being an essential and controversial part of the labor structure (compare Mintz 1960 and Taller de Formación Política 1982). Given the abundance of work in the cane, said Santos, why he decided to fish was "an interesting question." He followed with a lengthy story. As he recalled, the *mayordomos'* treatment of the labor force stimulated his break from agriculture. The catalytic moment came one day when those "*regando abono*" (who occupied the more highly skilled positions of fertilizing the fields) were forced to load wagons with sugarcane. This rupture in the division of labor, which involved the assignment of a harsher task, infuriated the workers, but for the most part they were obliged to comply; only Santos and a few others refused and were sent home for the rest of the day. The following day, Santos and the rest of the dissenters were assigned the harsh and dirty task of clearing the fields with a hoe. Santos recalls:

> I went to clear the fields with the hoe and returned back home tired, dirty, and all in sweat. *La doña* [the wife] then asked me what is worst [to fertilize the fields or to use the hoe]. [I replied], "Both!"

"Santos," she told me, "do not please the *mayordomos.*
Borrow a boat from *fulano* Miguel [so and so], and go fishing, since you have luck with it."

In 1955, Santos "became independent from sugarcane" when
he bought a small boat, a motor, and fifteen traps. He founded a
fledgling fishing operation that, with the help of members of his
household, enabled him to gain at least temporary independence.

Santos was not alone. Another Vieques fisher named Victor
became disengaged from sugarcane production through a similar
route. Yet Victor recognizes the difficult path of Vieques at the
mercy of absentee and local agricultural capital and pressures
from the U.S. Navy to obtain land for the development of bases,
target areas, and the deployment of troops and fleet. His household of origin was basically semiproletarianized, combining peasant production with casual farm labor. In his own words, "The
old man was a farmer more than anything else. He had a farm,
and we lived as *agregados* [laborer residents in an hacienda] on
the Benítez farm; he cultivated pigeon peas, eggplant, yams, sugarcane, and had cattle."

Around 1948, Victor worked in the cane, but at age eighteen
he managed to go in on some fishing traps with his brother.
In 1951, when he got married, he remembers that there was
chronic unemployment in Vieques. At the end of the harvest, *la bruja* caused a tense atmosphere on the island,
where people were trying to make ends meet with great difficulty. Victor worked cutting cane and *jalando azada* (pulling
a hoe) in the fields, but during *la bruja* he, his wife, and his
affinal relatives went into the hills to produce charcoal, and
Victor spent his spare time fishing.

Victor had bigger dreams for himself in the sugarcane
industry and tried to get a better position as a trucker. As
with Santos's experience, this decision came after an incident
in which those who cleared the fields were forced to work in
the pineapple harvest, difficult and hazardous work that
earned them less money. In 1952, Victor obtained his truck
driver's license, which allowed him to work carrying the cane
from the fields to the southern pier of La Esperanza, where

the sugarcane was shipped to the mills of Fajardo on Puerto Rico's East Coast. During the harvest, he transported cane, but during *la bruja,* he worked for the Commonwealth Land Authority clearing fields. With three of his brothers, Victor bought a small boat used for fishing close to the shore, which he used mainly for food and supplemental income. In 1958, Victor had a major argument with a *mayordomo* who treated the truck drivers with even less respect than he treated the cane cutters. That day, Victor quit the sugarcane industry, but he was able to support his family by fishing and driving a *carro público* (public transportation) in Vieques.

Leaving sugarcane production was not an easy decision. According to Victor, the early 1950s was a period of "great unemployment in Vieques." The employment situation in Vieques, once "a treasure" for job seekers, turned into a nightmare. Labor flowed to Vieques from the main island of Puerto Rico, from Culebra, and from the U.S. Virgin Islands, despite the bloody past of sugarcane strikes on the island (Meléndez 1986:19). Sugarcane production was on the rise; in fact, in the year of the Great Depression, 1932, the American Sugar Company, represented by the Fajardo Sugar Company, had reported profits of 300 percent. In addition to the world market conditions that accelerated the demise of sugar, in the case of Vieques, land committed to agricultural production was threatened by Public Law 528, which allowed the United States to establish military bases on Puerto Rican soil. This military conversion of land used to produce food, just as new immigration created a demand for fisheries products and small-scale-produced foodstuffs, was particularly opprobrious.

Yet it was wartime. In 1941, Public Law Number 54 conceded to the U.S. Navy vast coastal areas in Ceiba, Fajardo, and Naguabo, areas from which Puerto Rican laborers migrated in great numbers to Vieques, San Juan, and the United States (Meléndez 1986:47–48). The U.S. Navy initiated a systematic process of expropriation and relocation of small landholders and others who inhabited land given to the military. Violence ensued. In *La batalla de Vieques,* Arturo Meléndez contends that the expropriation of property that belonged to small landholders, along with the destruction of their farms and livestock, left them with no other choice than to become wage laborers in the construction of the military base—temporary employment that ended in 1948, starting "the via-crucis of the Viequense people: unemployment" (1986:50). In a letter to Governor Rexford Guy Tugwell published in *El Mundo* circa 1948, the testimony of Juan Nevares, a *viequense,* related how sugarcane land became idle federal land in the hands of the U.S. Navy, leaving ten thousand people to live in frightening misery. Nevares offered a list of recommendations for the improvement of the economy of Vieques, which included the transfer of funds and technology from the Department of Agriculture for the development of the local fisheries.

Meléndez (1986) describes the futile political maneuvers and negotiations of the government of Puerto Rico as it attempted to stop the U.S. Navy from expropriating Vieques by allocating other valuable coastal lands along the southern and northern coast of the island of Puerto Rico. Meanwhile, the mayor of Vieques, Antonio Rivera, constantly complained about the behavior of the marines in town and about the poverty of the *viequenses,* whose domestic animals died of hunger and thirst (1986:101–2). Land accumulation by a nonproductive enterprise left the people of Vieques homeless and at the mercy of the U.S. Navy and the commonwealth. Both entities continued to destroy the sugarcane industry, eventually forcing inhabitants into settlements in the center of the island, while two-thirds of the territory remained occupied by the U.S. Navy. The flow of people began to turn away from Vieques. In the 1950s, the population of Vieques started to migrate to Puerto Rico and to St. Croix in the U.S. Virgin Islands, to which Puerto Ricans had migrated as early as 1930 as sugarcane laborers and today form part of the island's fishing population.

## The Long and Winding Road of Deproletarianization

Among Puerto Ricans, as among most individuals who have joined the ranks of the unemployed and small-scale producers during modern times, the road of deproletarianization is a road less traveled. In the absence of a windfall, it is usually not an easy journey but one that requires many years of preparation, as well as migration among one's homeland, the U.S. mainland, and other locations both in and out of Puerto Rico. Although Santos's break with agriculture was virtually immediate, it took him nearly twenty years from the moment he initiated his path of semiproletarianization by learning how to use fishing to support his family even during *la bruja*. Santos was twenty-two years old when he started to fish, but he shared his household time and efforts between wage labor and fishing; the final transition to what he refers to as independence took two decades. Victor's transition—which involved other economic sectors and activities, such as public transportation and the operation of a small provision store—took half that time. By 1974, he had achieved independent production as a fisher/taxi driver who owned his own fishing gear and his own car.

Our third story of deproletarianization is that of Papotín, who operates a government-built *villa pesquera* at a public beach in Boquerón, on the Southwest Coast. He shares the operation with his current wife, a U.S. continental:

Papotín was born in Mayagüez, but in 1939, when he was seven years old, his mother separated from his father and moved to Boquerón, Cabo Rojo, where Papotín learned to fish. The local fishers taught him how to use trot lines and cast nets and how to collect mangrove oysters (*Cassostrea rizophorae*), which he sold on the street to tourists. Together, these activities provided food for Papotín and his mother through World War II. Papotín remained in school until the fifth grade, at which time he went to work in the sugarcane fields because his mother could no longer support the two of them by herself. From 1940 to 1950, Papotín worked sporadically as a cane cutter for twelve dollars a week, but he also always fished and collected oysters to help his mother.

From 1954 to 1957, Papotín served in the U.S. Army. Then he returned to Mayagüez to marry. During his first years of marriage, he worked as a laborer in a quarry and stone-products factory in the mornings and fished at night. Papotín engaged in this pattern of semiproletarianization until, coincidentally, the end of his marriage, in 1962. He believes that his divorce greatly affected his next step: migration to New York. In Brooklyn, Papotín spent a decade working as a machine operator in the garment industry, and he also remarried. In 1972, he returned to Puerto Rico with his wife and her children. Papotín and his wife had garment-factory jobs in Mayagüez until the factory closed in 1974. At this time, Papotín, now in his fifties, finally decided to fish full time.

Papotín's story illustrates the classic journey of Puerto Rican workers during the second half of the twentieth century. As with other sugar-dominated municipalities, Cabo Rojo, where Papotín worked as a cane cutter, was dramatically transformed by the industry. Sugarcane became the dominant crop and turned the area into a monocrop enclave. From 1900 to 1970, the proportion of land devoted to food crops decreased inversely with the increase in the proportion of land devoted to sugarcane (Valdés Pizzini 1985:65; Wessman 1977). The rise in sugarcane cultivation with the economic and political penetration of the United States also seems correlated with an increase in the population of Cabo Rojo throughout the first half of the twentieth century. Similarly, a period of population decline, from 1950 to 1960, coincided with the decline in coffee, sugar, and food crops. This was also the period of massive migration of Puerto Ricans to the United States and the onset of industrialization. In 1982, the return of Puerto Ricans who had migrated to the U.S. mainland contributed to a significant increase in the population of Cabo Rojo, and the area continued to grow through the 1980s and 1990s with the increased interest in southwestern Puerto Rico, particularly the municipalities of Lajas and Cabo Rojo, as tourist destinations.

During the 1940s, when Papotín engaged in cane cutting as a livelihood, production was at its peak in Cabo Rojo. Nearly 78 percent of all cultivated land was devoted to sugarcane (Picó 1983). During the next decade, cane land in the municipality declined, but Papotín took the path that carried many Puerto Ricans to the Korean War. Home from war in the 1960s, Papotín joined the ranks of laborers who were leaving the fields and moved to the U.S. mainland in search of a job and better living conditions. Shortly thereafter, however, the decline of urban economic development in the northeast cities of the United States, the region's fiscal crises, housing problems, and general industrial decline coincided with an expansion of the welfare state and the massive transfer of federal funds and aid to the poor in Puerto Rico (Castells 1984; Pratts 1987). Puerto Ricans such as Papotín, who had traditionally migrated to cities throughout New Jersey, New York, and New England, were among the first to be displaced, once again, as a result of economic decline. Both this decline and the expansion of public assistance fueled return migration to the island, and Papotín joined the exodus. Once in Puerto Rico, he made another attempt at proletarianization, taking a job in the garment industry in Mayagüez, a trade he had learned in New York. This time, Puerto Rico's tax exemptions were not enough to keep the factories open. Again displaced, Papotín finally turned to fishing, disengaging himself from the wage-labor path that had started during the same childhood years in which he had learned to rely on fishing. This time, following a route of "eternal recurrence," Papotín finally found his niche—where else but in his place of origin, Boquerón.

Elipidio Seda was from Arroyo, an area on the southern coast affected by high unemployment. Basically a proletarian, he also followed the long and winding road of deproletarianization. From 1948 to 1955, Elipidio worked loading sugarcane onto wagons in the Arroyo Central Azucarera. The next year, Elipidio began what would be a long job history, which included work on cauliflower farms in New York, migrating consecutively until 1959. In Puerto Rico, Elipidio worked in the building trade, mainly on the construction of the Phillips petrochemical plant in Guayama. The construction industry was at its peak in the 1960s, mostly because of an increase in housing projects and the building of the industrial infrastructure. With the help of affinal relatives and cousins, Elipidio was able to find jobs in that sector. With a *compadre*, Elipidio managed to get work doing masonry on housing projects in Guayama. After the construction boom ended locally, Elipidio traveled to Massachusetts to work for six months on tobacco farms, a job that he had found through the local office of the Department of Labor. Back in Puerto Rico, Elipidio worked in an auto parts plant and in a hospital, for a total of nine months. Then he went to New Jersey to work in a light-fixture factory. The labyrinthine journey of proletarianization in agriculture, the auto industry, farm labor in the United States, and construction finally came to an end in 1976, when Elipidio started to fish, this time for therapeutic reasons.

In these particular cases, fishing did not emerge as an immediate alternative to wagework in the sugarcane fields. All of these workers remained wage laborers for as long as they could. The benefits of social security and health and disability insurance and the security of weekly earnings could be neither denied nor downplayed. In some cases, fish market conditions (ex-vessel prices for the fishers), particularly during 1940 and 1950, were not lucrative enough to encourage full-time fishing as an economic alternative to wage labor. As the following section discusses in more detail, shifting from wagework to full-time fishing required the concerted efforts of the household and network members. Apparently, fish prices in the 1960s, when unemployment was at its greatest, were high enough to allow the transition to full-time fishing with the help of one's spouse and children and, in some cases, with the help of other (usually related) households in the community and spotty assistance from the state. Social security, food stamps, and welfare payments, made available to many of the fishers interviewed—especially late in their proletarian productive lives (and to some, injured or disabled on the job, in the absolute decline of their official proletarian lives)—assisted in the transition to full-time fishing.

During this time, these fishers found themselves in the same predicament as Taso and the social actors of *The People of Puerto Rico:* involved in sugarcane production because of the economic pervasiveness of the industry on the coastal plains, thus transformed into a rural proletariat, and suffering from a lack of similar economic alternatives along the coast or in the highlands (Picó 1983). Peasants, coffee and tobacco farmers from piedmont and highland areas, displaced by the U.S. tobacco companies, flocked to the coastal plains and urban centers in a desperate search for jobs or a stepping-stone to U.S. migration. The route of proletarianization that these rural households traveled took them to the sugar-cane fields, to farms on the mainland, to industrial and service jobs in the United States and in Puerto Rico, and (in some cases) to jobs provided by the state as subsidiary to the rural proletariat and as a way of maintaining a labor reserve (compare Scott 1976).

We noted in Chapter 2 that fisheries and the state played important roles in maintaining the labor force while severely dislocating economic and political processes were taking shape. At the same time, several acts of protest and resistance by coastal Puerto Ricans, including some with fishers at the forefront, accompanied the deepening social problems associated with the decline of traditional agricultural mainstays of the economy, such as coffee, tobacco, and (most notably along the coast) sugar.

During this difficult time, the critical importance of the island's fisheries cannot be minimized. It is the context in which coastal peoples, in vast numbers, derived solace after labor exploitation, therapy from injury or mental discomfort, food for their tables, and cash for their expenses, adding productive efforts to fisheries consistently used by a number of independent, specialized fishers. Independent producers, proletarians, and semi-proletarianized households used the marine resources in unison, each with its own schedule, productive arrangements, labor input, household commitments, and political praxis and discourse. Construction and industrial development—which inevitably proletarianize fishers or deproletarianize laborers and send them on the journey to their Ithaca—are also constant forces in the privatization of access to the shoreline, mangrove forest destruction, pollution, the systematic destruction of estuaries and nursery areas, and the perceived decline of fishing, thus making the pursuit of fishing as a full-time activity that much more difficult.

## THE HOUSEHOLD AS THE CONTEXT AND DOMAIN FOR DEPROLETARIANIZATION

In many instances, disengagement from wage labor required the full participation of household members. At the household level, moreover, the participation of the state cannot be overlooked, especially in the case of Puerto Rico, where the government participates in the economy as an agent of change and development in an eminently neocolonial, peripheral manner. Here we explain how the household provides the labor and support necessary for wageworkers to separate from their jobs en route to independent production. This analysis also examines the role of the state in frustrating or accelerating that process. Again, we look at the fishers who initiated their proletarian lives in the sugarcane fields. First we revisit the house of Victor:

## LA CASA DE VICTOR

Although trap fishing is Victor's main fishing endeavor, he uses cast nets and hand lines while his traps soak. His house, located on a hill with a view of the southern coast of Vieques, offers immediate evidence of his attachment to trap technology and his work space. Everywhere in Victor's yard are objects: cars, motor parts, wire, traps, and buoys. To enter his house, one must climb a short slope, traversing Victor's station wagon, his wife's small truck, a hog corral, cars used for spare parts, traps, and his sniffing and barking dogs. Before reaching the stairway of his cement-block and concrete house, one encounters an open space in which cast nets hang, traps and outboard motors rest, and two large wooden tables—used for fish cleaning, processing, and cooking—occupy center stage.

Inside the house is a small, comfortable, simple living room decorated with political posters. The kitchen is a large space that gives the impression of an industrial cookroom, with huge steel tables and three freezers. Victor is married to Eulalia, and they share the house with their son Ivan. Victor and Eulalia have five children, four of whom live in their own homes. However, their children's households form a network that plays an important role in Victor and Eulalia's objective of achieving independent production and deproletarianization. But perhaps more important, they comprise, to varying degrees, a dense network of households involved in the political and social issues that affect Vieques.

Prior to offering a formal description of Victor's and Eulalia's family network, it is pertinent to return once again to 1958, the year that Victor broke his ties with wage labor in the sugarcane fields. We have established that Victor left the industry because of problems with the *mayordomos*, but we also noted the difficult context of labor problems and economic difficulties in Vieques during that period (difficulties that still persist) and how Victor managed to survive *la bruja* by producing charcoal with Eulalia's relatives. Victor's transition also followed a long, winding road: His total reinsertion into fishing and independent production was not immediate; Victor drove his *público* (taxi) in Vieques from 1958 to 1974. Here we put this path into perspective. Victor worked for the owner of the *público*, who had moved to the Puerto Rican town of Arecibo. Victor kept 40 percent of all earnings and sent the remainder to the taxi owner. In 1960, with a one thousand–dollar loan from the local bank in Vieques and four hundred dollars in savings, Victor bought the *público*. Until 1974, Victor drove the car at his leisure, fishing at the end of each day of driving. During this time, Victor became president of the Unión de Choferes de Vieques (Vieques Chauffers' Union), the local syndicate for the *porteadores*. During his presidency, Victor negotiated a service contract with the U.S. Navy for the *porteadores* to provide transportation for visitors and personnel. The navy subsequently became one of the major employers in Vieques, a development that divided the people of Vieques into those who were for and those who were against the naval presence.

In 1974, Héctor, one of Victor's sons, found a job as a field laborer, and Victor allowed him to use his *público* to travel to and from work. In 1975, Victor shifted his entrepreneurial career in a new direction by establishing a *colmadito*, a small grocery store, while he continued to fish and sell his catch. (*Colmaditos* often become central spots where political and other issues are raised, argued, and discussed and where those who are down on their luck may be extended credit during difficult times.)

Victor's entrepreneurial path into the petty bourgeoisie is perhaps better understood in terms of the processes of capitalist differentiation among peasants. Fishers who produce a surplus often invest in small or limited trading activities, situating themselves into merchant capitalism. It is not uncommon for a Puerto Rican small-scale fisher to establish a *colmadito* to produce profits and cash for the household and to serve as an outlet for his or her catch. In some cases, this leads to the establishment of a *nevera*, or fish house (Valdés Pizzini 1985:273–75 [after Roseberry 1976]). Victor did not turn into a merchant or dealer who bought fish from other producers; he kept a market outlet for his own fish while he sold other goods and foodstuffs that he bought wholesale.

Several factors explain why Victor did not turn into a dealer. First, the fish market in Vieques was dominated by dealers from the main island, who traveled to Vieques daily on the Vieques-Fajardo ferry to buy fish in the harbor. Fish marketing then involved and still involves constant travel to and from Fajardo and the San Juan metropolitan area. Second, Victor was a true believer in cooperatives and associations (as evidenced by his presidency in the chauffeurs' union), and he became active in the local fishing association, in which he served at different times as president and vice president. A key objective of these associations, as we noted in earlier chapters, has been to provide a marketing mechanism for the local independent producers (although they can also become instruments for the formation of merchant capital). As a result, these associations can protect members from the business practices of dealers and provide a basis for political organization for the protection of the community of fishers and the community at large (compare Valdés Pizzini 1990b).

Victor, Eulalia, and their son Ivan, age 24, live in the house they own near Barrio la PRAA, a community named after the federal project that built houses in the area. Victor and Eulalia's purchase of their home came as a result of various lifetime permutations. Victor, who was born in Punta Arena on the northern coast of Vieques, moved to El Destino, where he now lives, in 1941, when the U.S. Navy expropriated the land where his father and family lived as *agregados* (small landowners). Instead of moving to Montesanto, a U.S. Navy–owned landholding *"donde hiban los expropias"* ("where the expropriated are relocated"), his father bought a piece of land in El Destino. According to Victor, his family refused to move to the landholding in Montesanto because, as U.S. Navy property, it could be claimed for other use at any time, and relocated families would be forced to leave once more.

During the hard times of 1952 and 1953, Victor and Eulalia managed to buy a house, which they paid off in one year. In 1957, Victor and Eulalia sold the house, along with several hogs and fighting cocks, to one of Victor's brothers for three hundred dollars. With three days of pay that they had managed to save, they bought their current house from a family that left Vieques for Fajardo. Victor maintains that the predicament of Vieques centered on a scarcity of economic opportunity, which fostered emigration to mainland Puerto Rico, the U.S. Virgin Islands, and the U.S. mainland. Yet in the eye of that economic and social turbulence, Victor's household invoked several economic and political strategies that allowed its members to stay in Vieques. As Victor expressed it, "[One] has to go to the sea as a solution. . . . To remain in Vieques, one has to get his hand in whatever is needed. I have the greatest pride in saying that I have not earned a cent outside of Vieques."

In contrast with the life trajectory of the majority of Puerto Rican fishers, Victor's life trajectory evidenced resisted migration and reincorporation into the proletarian status through an increase in time spent as an independent producer and merchant. This did not prevent him, however, from keeping an eye out for wagework that he might enjoy. In 1972, after he learned that the commonwealth's Port Authority had vacancies for sailors on the Fajardo-Vieques ferry, Victor applied for a position. But according to Victor, a change in government (the PDP's return to power) prevented him from getting the job. In his own words: "Small town, big bell [everyone knows you and knows your party of affiliation]. If you do not take communion [with the political parties], you do not eat.

I'd rather eat dirt." Although Victor is not typical of a committed independent producer, a laborer deproletarianized, it is also clear that the alternative of reincorporation into wage labor—especially in low-remuneration, low-skill, and low-workload jobs in government-related activities—can be a lucrative option for fishers. These jobs offer health insurance, retirement plans, and other benefits in exchange for minimal effort and a schedule that allows continued fishing. Some of the Puerto Ricans we interviewed in the summer of 2000 reported that their principal reason for leaving independent work, such as *chiripas,* for wagework, was to receive social security benefits. Victor shares with his many peers the therapy explanation for his involvement in fishing: "When people go to the sea, one forgets the problems, and breathes fresh air; one forgets everything. . . . Fishing is a therapy. When I was younger [a cane cutter], I used to drink from the start of the weekend to its very end. Then I could be dying but made an effort and went to the sea and became a new man recuperated from the ills of drinking. I leave all impurities in the sea."

A lifetime perspective of Victor's personal trajectory as a deproletarianized sugarcane worker includes his participation in an array of economic activities. Here we examine the details of his lifetime strategies, especially the ways that other members of his household figured into his difficulties. Victor's wife, Eulalia, has been the hub of those strategies. Eulalia works as a homemaker but also has a *guagüita* (a van), which she uses on weekends to sell fritters to tourists and visitors in the principal city of Vieques, La Esperanza. Once the harbor of a combative fishing association, La Esperanza has been gentrified into a barrio that shares shoreline and coastal plain with *parcelas* (rural settlements); a *parador* (hotel); and a multitude of businesses, restaurants, bars, and guest houses (most owned by people from the U.S. mainland. La Esperanza, which used to attract reporters to cover the events of the fisher resistance and opposition to and interruption of the U.S. Navy target practices, now attracts Puerto Rican tourists in the summer and foreigners and people from the U.S. mainland in autumn and winter. Eulalia makes *empanadillas* from conch (*Strombus gigas*) and trunkfish (*Lactrophrys trigonus*). Victor traps the trunkfish and Manuel, one of Victor and Eulalia's sons who lives in La Esperanza, provides his mother with conch. Eulalia also prepares cakes for weddings and special occasions, weaves tablecloths, and sews clothes for women. After describing Eulalia's skills, Victor declared, "*Suerte a eso*" ("Luckily for that" or "Thanks to that"), a profound phrase that defies true translation. *Suerte a eso* is used to signify that an activity or object makes a difference in the outcome of an event or even makes an event possible. *Suerto a* Eulalia, the household and network of related households have maintained their economic integrity. Usually, in fishing households women help clean the fish. In this household, in contrast, Victor helps Eulalia clean and prepare the conch and trunkfish.

## THE HOUSEHOLD NETWORK AS A CONTAINER OF FISHERS OR AS A CONTAINER OF LABORERS?

Households are, in the context of fishing, containers, or units, that engender and maintain fishers in their communities and cushion them from tensions produced by mercantilism and other entangling capitalist relations (see Sider 1986). Still, fishers cannot escape these tensions, whether they are produced by merchant capital within or other forms of capital generated outside of fishers' communities. In Puerto Rico, the logic and structure of the economic system allow each household to coreproduce, among other commodities, laborers for industry, for public service, or for commerce. As a result, fishing households in Puerto Rico, perhaps more than anywhere else, experience strong tension in the socialization of children with respect to the skills and productive culture of fishing at the same time that the state and its expectations pull them toward proletarian status. Further, there is always the alternative choice of joining the ranks of migrant workers who board planes to the United States. Eulalia and Victor's household experiences these intersections and tensions with capital; as a result, the household becomes a container of laborers, despite its petty-commodity character.

When we interviewed Eulalia and Victor, as we have mentioned, their household comprised them and their youngest son, Ivan. Ivan helped Victor with the fishing chores. Ivan was also a skilled CB and heavy equipment operator and had worked briefly for a government agency in Vieques. Perhaps in solidarity with his father's objectives, however, Ivan did not follow the agency when it moved to the island of Culebra. Instead, he remained in Vieques to help his father and perform *chiripas,* thus establishing himself in the trajectory of semi-proletarianization. Carlos Rubén, another son, also helps his father with the fishing. Although Carlos has a teacher's certificate in biology and taught at the town's high school, according to Victor, "he has to fish, since there is no work for him." His wife, who has a bachelor's degree in nursing, is also unemployed. During our fieldwork, she was applying for a position at one of the few factories in Vieques. Cheito, another son, had a provision store in La Esperanza, and Junior, yet another son, has been both a fireman and a fisherman. Ángel Manuel, still another son, worked for several years in a drugstore in town. When he was forced to leave that job, his father helped him begin to fish. Ángel Manuel provided his mother with conch and fish for her business.

Thus this network of interrelated households that originated with Eulalia's and Victor's house—despite the original household's petty-commodity/capitalist character—becomes a container of laborers. And perhaps more in line with the arguments elaborated by Meillasoux (1972), it becomes a unit that supports and contributes to the reproduction of the labor force during the peaks and valleys of capitalism. Obviously, Victor and Eulalia's children have the option of migrating, but their resistance against doing so; their personal and political commitment to the welfare of Vieques; and the powerful economic strength of the network of household businesses, labor pool, infrastructure, and productive means allow them to continue in a state of noncommoditization of their labor, in the hearth of independent production.

The share system that Victor uses illustrates this point. After Victor sells his catch and subtracts expenses, he divides the money into three equal parts, even though he owns the boat and the gear. In return, his sons help him with maintenance, motor repair, woodcutting for the trap frames (a job that Eulalia also helps with occasionally), and net mending. However, if his sons have loans to pay or other extra expenses, Victor waives the share system and gives them the money they need. As a result, their ability to cover basic financial needs and expenses is not circumscribed by how much they fish; they know that the cash will be provided, financed by their father's fishing and his *colmadito* and, perhaps most important, by their mother's fritter and cake business in La Esperanza. Each household, with its own schedule and strategy of production, offers economic asylum to any member who becomes unemployed, providing the means to support the individual's family until his or her situation improves.

## A LIFETIME OF LABOR, INDEPENDENT PRODUCTION, AND POLITICAL PRAXIS

An exploration of the specifics of the trajectory that allowed Victor to participate in wage labor to become a full-time fisher is now in order. In Mintz's (1956) discussion of fishing in Cañamelar, he observed that becoming a full-time fisher presented insurmountable hurdles for the rural proletariat. The problem was that the cost of sailboats, which represented a significant accumulation of capital, was prohibitive for the cane workers. The time that wage laborers committed to sugarcane production also precluded their participation in full-time fishing. Thus, fishing was relegated to a subsidiary, seasonal position, as an activity that was practiced primarily during *la bruja*. Low prices for fish, lack of refrigeration facilities, and the required capital investment and reinvestment made fishing a difficult trade to enter. Government officials, evaluating the fisheries early in the twentieth century, agreed that, in the context of the prices and marketing structure, the fishers' earnings were so meager that they did not allow for an adequate subsistence or for the required intensive investment of capital (Jarvis 1932; Vélez, Díaz Pacheco, and Vázquez Calcerrada 1945).

Victor's life history sheds light on these difficulties. He started fishing with a brother who had a boat. They paid a man to build traps and started their career in fishing. Later they pooled their money to buy a small boat. Victor's share came from his wage labor in the cane. According to Victor, their labor commitments allowed them only to "fish close to the shore," basically to eat but with some left over to sell. In 1953, Victor bought a ten-horsepower outboard motor, financed by the Department of Agriculture. This event is significant. The postwar period marked the decline of sugarcane on the island, in response to world prices, and the emphasis on industrial development and changes in the patterns of land use. It also witnessed an increase in the prices of fish and shellfish. Coincidentally, government programs for the reincorporation of labor into other sectors of the economy and the support of sugarcane laborers appeared jointly with efforts to "develop" the local artisanal fisheries.

In line with the strategy for development, the commonwealth initiated efforts to increase production, improve marketing facilities, and expand fishers' opportunities at the same time that prices increased (Iñigo 1968). Government efforts included the motorization of the fleet through the granting of loans for their purchase (this created the fishing credit section at the Department of Agriculture), centers for the sale of low-cost gear and equipment, fisher training courses, and the construction of *villas pesqueras.* Victor, like many of the fishers we interviewed, bought motors and boats only when the capital became available through government development programs. Victor's peers also benefited from this situation.

Yet at Eulalia and Victor's house, success in fishing and the accumulation of capital depended on the array of activities in which both engaged: Eulalia's selling cakes and fritters and Victor's working in the cane, driving the *público,* and selling food-stuffs at his *colmadito.* Although Victor had *proeles* from outside his household network (including his brothers), his success came primarily from endofamiliar accumulation or capital generated by family labor, often unremunerated, mainly performed by his sons (Cook 1982). Money transfers from various activities supported other endeavors. For example, Victor acquired loans to buy the *público* and a ten thousand–dollar bank loan to buy a fishing boat in the early 1980s. Around 1958, Victor had a *siembra* (small farm) with various crops, which he sold to support his family. He recalls that he was sometimes forced to sell cows to pay for the *público*, a common strategy among farmers worldwide who have trouble meeting large household or business expenses.

Life in the household of Eulalia and Victor was also marked by periods of intense political activity that became intertwined with the process of capital accumulation and household labor. As we mentioned, Victor had experience with the Unión de Choferes de Vieques, negotiating with the government; the public; and most important, with *La Marina* (the U.S. Navy), an agency with which there was much contention and conflict in Vieques. Victor, who joined the local fishing association, declares that he "always believed in unions, cooperatives." "I know for a fact that if there is *compañerismo* [camaraderie], they could be successful," he says. But Victor also joined the association to become incorporated in a formal organization that might allow local fishers to break the price and acquisition monopoly of seafood dealers from the main island of Puerto Rico, to improve the infrastructure, and to obtain loans for gear and boats. In that context, the availability of fish was crucial, but it inevitably caused conflict between fishers and the naval base over access to certain fishing areas and the destruction of nursery habitats, where certain species grew, spawned, and lived. Much of the land expropriated by the U.S. Navy was made up of mangrove forests, coastal lagoons, and bays, areas suitable for fishing and stock reproduction.

In the second half of the 1970s, the Asociación Pesquera de Vieques, one of the first of its kind, benefited from the development programs of the commonwealth's Acción Comunal (Community Action), which was committed to the developing Puerto Rican fisheries. A typical Acción Comunal project included the refurbishing of landing infrastructure and the transfer of boats to the association. Although some of the boats purchased were either trawlers or large boats that were unsuitable for tropical waters because of the amount of fuel they required for operation, Acción Comunal also provided the association with personnel, including a secretary, an administrator, and a salesperson. In addition, Acción Comunal provided information transfer of navigation technology to the local fishers. To operate the boats, Acción Comunal hired captains (Victor was hired as captain of La Esperanza I) and crew. In most fishing communities, the project failed, since, ironically, fishers who were hired as wageworkers no longer had to fish to earn a living.

Victor and his sons actively participated in the association, especially in 1978, when the fishers challenged *La Marina* by invading its bombing ranges and other areas of military maneuvers, forcing the ships and cutters to chase the fishers' boats. Victor and his family vehemently opposed *La Marina*, which they considered the cause of the destruction of habitats and the snapper and bait fish fishery in Vieques. The loss of traps and the creation of ghost traps as a result of the constant crossing of U.S. vessels have forced most trap fishers to abandon the gear. As Victor explains, "*La Marina* continues in Vieques. Destruction of the *arte* [fishing gear] continues, a cause for the extermination [of fish]. The U.S. Navy boats cut the rope, and the traps remain killing fish."

During these ongoing disputes, Victor acquired a wealth of experience in community organization and political expertise. The involvement of leftist movements, parties, and organizations—those who advanced Puerto Rican independence—pushed him in that direction. "We were involved with the Left in this country," recalls Victor with mixed feelings. The incorporation of the Left was expected, since Vieques is part of "the struggle against colonialism and US imperialism" (Meléndez 1986). Since 1930, the pro-independence movement and its variants, some of which practice terrorism, have been involved in the struggle of Vieques, having successfully achieved demilitarization of the nearby island of Culebra. This activity continued through the end of the twentieth century and into the twenty-first century (see Chapter 8).

The training of fishers in the fight against imperialism took them to Caribbean countries, such as Grenada, where they observed fisheries development and met Maurice Bishop, as Victor did. Although Victor acknowledged the importance of the cadre of intellectuals and leftist organizers, he and his sons sadly report that the domination of the movement against *La Marina* by leftist organizations alienated the majority of the *viequenses*. Unfortunately, our only information about the association is what

Victor and his sons shared with us. They told us that they did everything to reach an agreement with the navy's admiral, to share schedules in the use of the fishing areas, and to allow the fishers to enter various areas. According to Victor, the leftist elements would not comply with the terms and managed to persuade many fishers to follow them. Eventually, the leftists succeeded only in isolating themselves.

Sadly, using a voice and rhetoric charged with emotion and frustration, Victor deconstructs the history of the heroes of the "struggle of Vieques." He becomes emotional because his side of the story is the side of those who, necessarily, created the schism in the movement, trying to save it for the sake of the *viequenses*, and he becomes frustrated because he recognizes that *La Marina* and all its problems still remain.

## HOUSEHOLDS IN PERSPECTIVE

The life history of Victor, Eulalia, and their household, singular though it is, provides us with the ingredients to construct a paradigm of household incorporation and intervention in the process of deproletarianization. It is clear that Victor's household's deep and enduring economic and political commitment has been responsible for its success in maintaining both independent production and a network of semiproletarianized households. Yet the participation of other household members in the fishing enterprise is crucial to their success as well, even in cases without large pools of household or network labor. Despite the fact that Papotín and Santos have few children and are not entirely devoted to fishing, they received timely and crucial assistance from their wives, making the transition from wage labor to fishing not only smoother but ultimately possible. Papotín's wife was also his business partner. Santos's second wife encouraged him to leave work in the cane: She, he says, "opened my eyes." A woman from the highlands in Puerto Rico, she may never have helped much in fishing chores; yet she managed the household with the financial precision necessary to withhold enough money from consumption to buy a boat.

Thus fishing households, like peasant households, are key elements in the trajectories outlined in this book. The degree and success of deproletarianization depends on household structure and composition; its ability to draw others into its fishing operation; and the utilization of joint, complementary skills in supporting these interrelated households. The state may facilitate this process, its role may be relatively benign, or it may frustrate the reproduction of fishing as a way of life. The system permits fishers to benefit from social security, food stamps, and disability benefits. Use of public education represents an attempt by both the household and the state to incorporate children into the university system, the armed forces, the flow of migrants to the United States, or the local workforce. Inevitably, they become only partially separated from fishing, learning enough of the trade to keep them occupied if they cannot find jobs or if they become unemployed.

## DREAMS OF INDEPENDENCE

The implication is that fishing subsidizes the development of other economic sectors, reproducing workers who can be attracted into the workforce during times of economic growth yet who can move back into fishing during economic declines without severe economic consequences to capital (for example, worker's compensation, unemployment insurance). Yet in journeys between fishing and wagework, most fisher/laborers intend to decommoditize their labor and become deproletarianized in the long run. Although the labor becomes commoditized in relation to several factors (for example, needs of capital, patterns of labor remuneration, structure of household participation in labor, political affiliations and strength, legal protection, and laborers' availability), Puerto Rican fishers resist this in whole or in part by fishing. They are well aware of the myriad crises that send laborers to the ranks of the unemployed, although not necessarily to the ranks of the idle and unproductive, and these constant reminders encourage them to dream of independence from the whims of capital and the difficulties of low-wage jobs. In conversations about their lives, Puerto Rican fishers, closely tied to the movements of the world economy, feature a desire to decommoditize their labor and become independent. The Puerto Rican migrant nostalgia (often expressed in the saying "New York is only a couple hundred dollars away"), sporadic unemployment, and personal or domestic problems all encourage the proletariat to daydream about independence from wage labor—whether in the midst of repetitive activities, in the loneliness of a ship on the Pacific, in the boredom of a factory in Mayagüez, amid fiberglass boat fumes in a New Jersey factory, or in a salt pond in Cabo Rojo.

In comparison with jobs in extractive industries and agriculture, industrial work—despite its potential for injury—generally offers higher pay and benefits to laborers, making it a preferred option among many Puerto Ricans. Industrial jobs, however, usually require workers who are semiskilled, skilled, and educated—a caveat that discourages and disqualifies many Puerto Rican fishers, especially those who left school as young children to enter the workforce. Our final profile presents a somewhat unusual scenario. It features a fisher who moved to the industrial sector, worked at what appeared to be a good job, and suddenly left this employment to become independent.

## Deproletarianization of a Skilled Laborer

Rudy Irizarry has adopted a multigear, multispecies approach to fishing, using both gill nets and traps as his principal gear. Evidence of his full-time commitment to fishing are his 120 traps, which he uses to fish three days a week and which he has devoted more than two days per week to maintaining (somewhat less after he began to use vinyl-coated wire for his traps). He also owns three gill nets, which he uses to fish in twilight when he is free from the traps.

Although fully prepared to enter the ranks of the industrial labor force as a skilled professional (a technician), Rudy has declined this opportunity. In contrast with most fishers, he was just two courses short of a bachelor's degree in medical technology when he turned to fishing. He explains:

I went to the university, but I used to fish then. As a matter of fact, most of my books and school materials were paid with the catch. I spent three and a half years

studying medical technology, with two courses remaining
for graduation, but I realized that *lo mio era la pesca* [my
calling was fishing], to be in the boats, in the sea, and I
left school. Neither laboratories nor the small bottles
caught my attention, nor being in an office. I started to
fish because *eso es lo mio,* and you can make a living out
of fishing. . . . First of all, I studied because, since childhood, my father inculcated [in me] that school was first
and then fishing. I went to grammar school, to high
school, and college, and he was happy. The day I told him
I left college, he almost killed me. With college, you can
have more opportunities and benefits. But it did not call
my attention. I like to fish.

Such pressures are easily understood against the labor background of Rudy's household and the community. The two families that form the axis of his kinship network are original settlers
from La Parguera, a coastal community whose population was
relocated to a local *parcela* with no access to the shore. Few
households from the original group of settlers were left with property along the shoreline. La Parguera's bioluminescent waters,
mangrove forests and canals, reefs, mangrove islets, and solitude
attracted middle-class settlers and tourists to the area. The new
settlers built stilt houses in the water, many of them illegally,
thus occupying most of the shoreline and the nearby islets. Today
more than four hundred such houses share community identification with the people of La Parguera (see Chapter 7). The influx
of people and capital soon impacted the local population, and
entrepreneurs and landowners from the town of Lajas divided the
land into real estate property and developed tourist businesses,
guest houses, hotels, and restaurants.

Thus, a community of rural workers in salt production, sugarcane, and cattle grazing; fishers; and peasants had to face the
process of coastal gentrification, in which "outsiders" displaced
the local population demographically, socially, and politically
(compare Valdés Pizzini, Gutiérrez Sánchez, and Chaparro 1988).
This commoditization of the coastal zone also stimulated efforts
to protect La Parguera's marine resources from commoditization

and use by means of a marine sanctuary. There was a massive movement against the sanctuary (discussed in detail in Chapter 7). It was a movement spearheaded primarily by local fishers with the help of fisher organizations in the western portion of the island and the recently created Island Fishermen's Congress (see Valdés Pizzini 1990b).

One of the largest local employers of fishers was the University of Puerto Rico, through its Graduate Department of Marine Sciences. This department has laboratories and classrooms on the islet of Magüeyes, the largest islet in front of the shoreline of La Parguera. Because the university has a physical and human presence in the community, with faculty, students, and La Parguera residents sharing living and leisure space, a close relationship has developed between the university and native communities. Rudy's family, one of several that secured access to the best jobs in the department, has a clear history of semiproletarianization. On one hand, they fish the reefs of La Parguera; on the other hand, they land jobs at the department as janitors, supervisors, mechanics, research assistants, and boat pilots, in addition to performing leisure-oriented *chiripas* such as repairing boats, ferrying, and maintaining sportfishing boats. This is the entanglement of expectations in which Rudy constantly finds himself: the route of semiproletarianization, the economic and social advantages of government work (which allows time for fishing), and the potential rewards of academic endeavors.

After the rupture with his household's expectations, Rudy entered the local industrial sector. A brother worked as a supervisor in a pharmaceutical plant in the city of Mayagüez, fifty minutes from La Parguera, and Rudy took a job there. After several interviews, we still do not fully comprehend his reasons for taking the job at the factory. We suspect that he was forced to take the job after he left the university, as a way of channeling his future efforts back to college, since he was appointed to work in the laboratory with those symbols of tedium—the small bottles. We do know that Rudy entered industrial work with hesitation and apprehension and always maintained to his brother and father that he would go back to fishing if he did not like it. Rudy claims that, because in wagework he has to pay social security and income tax, he earns more money fishing. He was also weary of the dilemma of working for himself or working for the government. He decided to become an independent producer, to become separated from wage labor and its particularities, a decision that required, in his view, a total time and effort commitment. In fact, for Rudy, "a commercial fisherman is one who fishes full-time. If he works somewhere else, he is just a laborer [*empleado*, or employee]."

### The Force of Household

Success in becoming independent from wage labor, as we have shown, usually depends on household participation in fishing activities, *chiripas* to sustain the family, and money transferred by the state.

> In Rudy's case, his wife and in-laws are crucial to their household's survival and well-being. Rudy lives in his in-laws' home, which includes plenty of work and storage space for his fishing gear and equipment and an old butcher shop that he uses now to sell fish. After working as a butcher, his father-in-law became an electrician and is now fully occupied with all the house construction and remodeling in La Parguera. Rudy's mother-in-law is a homemaker. Both of his wife's parents help him with the transportation of fishing gear, boat and motor repair, and fish cleaning.
>
> His father-in-law and his mother-in-law both come from households involved in small businesses. This is an asset to Rudy, who admits that he lacks business skills, as well as an asset to all household members who wish to remain independent from wage labor. Ana, Rudy's wife, who has a degree in nursing, was unable to get a job in a local hospital. Instead, she worked briefly as a schoolteacher. Having little luck in the formal economy, Ana became an artisan. She works with various marine materials, such as tarpon scales, which she transforms into earrings, bracelets, flower arrangements, necklaces, and souvenirs. In addition to helping Rudy clean the fish, Ana's mother has become a working partner in her daughter's art-and-crafts business. Buyers come

directly to their home, but each weekend Ana greets the tourists from a table at the harbor, where they flock to board boats to the bioluminescent bay. Entrepreneurship has paid off for Ana; it has financed a sport utility vehicle to carry her crafts and materials. The purchase of the Irizarrys' car on credit and their account at the local cooperative have helped Rudy get loans to buy motors and a boat. It would be difficult to say whether Ana's or her husband's endeavors are more profitable: Rudy is one of the most successful fishers not only in Lajas but in all of Puerto Rico; yet Ana's business is quite profitable as well. A successful and recognized artisan, Ana has been honored by the Office of Economic Development for her achievement, and she is a member of the artisan registry of the Institute for Puerto Rican Culture. Her official recognition as an artisan has produced a stream of invitations to present and sell her crafts in different state-sponsored cultural activities.

Ana's success is evident too in the couple's use of domestic space. Although Rudy has a fish house at home, he sells most of his catch to a friend because he does not have the time to clean, prepare, and market the fish and because he feels that he does not have a business mind as his wife and in-laws do. Nevertheless, he keeps fresh fish in the fish house, which is regularly patronized (his in-laws and his wife sell the fish when he is out). Now that his wife is doing well with the business, it is hard to walk into the petit fish house without stumbling over boxes of raw materials and processed shells for sale. It is clear that she has invaded a portion of what used to be her husband's work space. "She moved into my fish house," says Rudy.

Ana's expanding business has also penetrated and altered her husband's work schedule and the direction of his efforts. Rudy admits that, during the summer, the pressure to procure materials is too great for him to bear alone. Sometimes, he says, "they [mother and daughter] fight with me" to get them materials. From her husband's incidental catch of sharks, Ana uses the teeth, and twice per week Rudy dives for shells, broken pieces of coral, and other marine materials for his wife's crafts. When he sells lobsters to the restaurants, he asks for their eyes, which Ana uses in her creations. And twice per week he searches for *bulgao*, a type of gastropod that congregates close to the mangrove islets, is good for eating, and is excellent for the crafts that Ana makes. Rudy also fishes for snook for its huge scales, which his wife uses. Sometimes Rudy hires a diver to bring back conchs and shells. Ana also purchases materials from a Florida warehouse that imports shells from the Philippines and the Pacific Ocean.

## A Potential Proletarian and the Politics of Labor and Fishing

Rudy's brief encounter with industrial work ended in a sharp separation from wage labor in order to fish. To make this transition, he relied heavily on endofamiliar accumulation, household labor provided by his in-laws, and his wife's craft business (which modifies his fishing schedule). During our fieldwork, Ana gave birth to a child, and with the help of her parents, she and her husband were able to continue to sell fish and crafts during her hospital stay. Like many Puerto Real fishers and merchants (see Chapter 3), Rudy seems to understand that capital accumulation in a fishing enterprise (or firm) requires the use of household members, a strong bond between father-in-law and son-in-law, and continued cash flow from fish sales. Consistent investment in equipment, boats, and gear are also key factors in the firm's success. In the face of diminished support from the federal government and few loans for fishers, Rudy does not depend on money transfers from the state. Instead, he obtains money from the local banking cooperative by using his wife's credit.

In more ways than one, Rudy defends the interests of all small-scale fishers who use the resources of the platform, particularly in the fight against fish dealers who contend that fishing permits for other islands and polities in the Caribbean are needed. From Rudy, dealers buy frozen fish, especially during Lent, the peak period for demand. Focusing on the snapper/grouper fishery in the shelf drop-off and in similar fish banks in the Caribbean, these fishers and the dealers who control their production withdraw from the politics of fishing (see Valdés Pizzini 1990b; see also

Chapter 7 of this volume). Rudy complains that the "fishers here do not care [about the resources]," and insists that the fishers "have to protect" their "product." For Rudy, fisheries management protects recreational and sportfishers at the expense of the commercial fishing livelihood, limiting access to marine resources.

Yet Rudy is not against conservation: "I am conscientious," he says. "I understand. I am all for conservation," but "both sides [fishers and fishery managers] have to be flexible." He sees that laws "are for some, but not for others." Clearly, this is a reference to the situation in La Parguera, where the owners of the stilt houses remain, violating both federal and commonwealth laws concerning mangrove destruction and unlawful construction of structures in waterways, while fishers trying to make a living are heavily monitored. How is it that the *caseteros* (stilt house owners) cut mangroves without any penalty? Rudy maintains that the DNR provides the permits and that its personnel, hired on a political basis, seem to prefer the owners of the *casetas* ("lawyers and physicians,") to the fishers. "If a fisherman had a *caseta* in the sea," he said, "the DNR would take the house away and burn it."

Rudy admits that the fishers have an impact (often destructive) on the environment and resources, but he sees outside pollution as a major problem as well. Yet the state prefers to harass the fishers rather than deal with the middle-class professionals who own the stilt houses in La Parguera. "The government," he claims, "is more corrupt than we are."

After examining Rudy's social, economic, and political trajectory, which is firmly established in independent production, it is difficult to visualize him as a wage laborer. What events could precipitate his departure into wage labor? According to Rudy, "an economic depression," a difficult situation at home, triggered perhaps by a fishing prohibition, could do so. But the return to (or reoccurrence of?) wage labor seems to be omnipresent in the fishers' minds, and so it is in Rudy's. After leaving the pharmaceutical plant, he applied for a job at the DNR, but, he says, "it was denied since my father was not from the Popular Democratic Party."

And during our fieldwork, Rudy did, indeed, apply for a job—where else?—at the Department of Marine Science. He wanted to work the shift from 4:00 A.M. until noon, which would allow him to fish part time. With a newborn in the family, the perceived threat of a fishery closure, and the stochastic character of fishing, he had to consider reentering wage labor; yet he did so in a way that permitted some fishing. He will remain a fisher; yet—as with his siblings, cousins, and uncles who work at Isla Magüeyes—he will also serve science and the state, a semiproletarian.

# 7    Power Games

## *Work Versus Leisure Along Puerto Rico's Coast*

IN THE LATE 1980s, working on behalf of the Southeast Regional Office of the National Marine Fisheries Service (NMFS), we and several colleagues spent the better part of a year visiting coastal municipalities and small islands in the Puerto Rico/U.S. Virgin Island archipelago, interviewing sportfishers and creating an inventory of recreational infrastructure. This work was part of the NMFS's attempt to incorporate into its research agenda more studies of leisure uses of the nation's coastal and marine resources, which constituted a shift away from its previous focus on commercial uses of the coast. These shifting research priorities reflected the growing political and economic influence of leisure capital interests on or near the coasts of the Americas. Through the 1990s, tourism became one of the largest global industries in the world, converting marine and coastal landscapes in ways that have far-reaching implications for the labor and livelihood of fishing families (Griffith 1999). Our work for NMFS involved documenting the existence of and discussing the use of marinas, launching ramps, boat storage areas, and the like; speaking with members of sportfishing clubs; discovering budding or incipient boating, lodging, and other tourist facilities; and determining the links between commercial and recreational uses of the coast.

Like coastal communities throughout the Americas, Puerto Rico's coastal towns have witnessed growth proceeding at rates far higher than most inland regions, creating new jobs and new business opportunities while threatening to exacerbate conflicts over the uses of public and private lands and waters. Although growth in urban port and harbor development, conventional tourism, ecotourism, heritage tourism, and development oriented

194

toward seasonal living and retirement create investment and employment opportunities, they also compete with one another for space, labor, capital, and political favor. These developments occur as commercial fishing and seafood processing, long main-stays of coastal environments, suffer from several problems related to ecosystem changes and new regulatory environments. These problems are likely to arrest the future growth of and alter the complexions of commercial fishing and related industries (Durrenberger 1992, 1995; Griffith 1997, 1999; Maril 1995).

One of the critical processes that triggers disagreements and conflicts between the environmental movements and the state and private sectors in Puerto Rico is the sharp increase in tourism and recreational infrastructure in the coastal zone. Most leisure and tourism growth has occurred along the West and East Coasts. Throughout the Caribbean, urban, tourist, and leisure development in the coastal zone tends to directly affect traditional fishing communities in negative ways (Stonich 1998). Puerto Rican fishers, who are directly affected by devel-opment and by conservation measures, have developed a criti-cal outlook and have taken an active political stance that puts them at the forefront of many environmental causes. In 1983, the National Oceanic and Atmospheric Administration (NOAA) Marine Sanctuary Program, attempting to develop a sanctuary for conservation and recreational purposes along the Southwest Coast, encountered fierce opposition by fishers and community organizations. Members of the local community and fishers themselves felt that a federal marine sanctuary would, in fact, curtail their freedom, penalizing them instead of confronting those principally responsible for mangrove destruction and sewage disposal: specifically, upper-class absentee owners of *hojes* (stilt houses) used as second residences, who had control over the shoreline. We examine this situation in detail later in this chapter.

Economic restructuring, with new political economic agen-das in competition with one another while traditional ones either remain stable or enter a decline, inevitably creates social problems among residents and between residents and new-comers. Few countries of the Americas have been spared. New regulations on groundfishing in the Northeast United States, cutting fishing incomes in half, have been enacted against a background of a more generalized decline in textile and other industrial production (Griffith and Dyer 1996). Economic dislocation has been even more common and more disruptive in other parts of the Americas (Nash 1994). Throughout the Caribbean and coastal Latin America, corporations such as Hilton and Marriott have privatized stretches of coastline for resorts that are far beyond the consumption levels of native inhabitants. Gentrification, changing environmental laws, and the strategic use of coastlines by Caribbean and Latin American military forces—in part justified by the trade in immigrants and drugs, both intimately tied to tourism—have displaced natives of coastal dwellings or threatened their ability to include the sea and its resources in their livelihoods, forcing shifts in economic and political strategies.

In previous chapters, we have seen that Puerto Rican fishers, faced with constricting job markets and onerous developments in working conditions, tend to move back into fishing as a form of therapy and a source of income. What happens when this option—the fishing option—also becomes restricted or altered in such ways that fishing no longer offers release from stress and a reliable source of income? Two disputes over coastal access in Puerto Rico raise this question in different ways; yet both illustrate how disputes between leisure capital and fishers draw upon essential characteristics of fishing communities. In the process, these disputes strengthen the organizational powers of fishers as they tax those powers and expose cracks of vulnerability within Puerto Rican fishing communities. The first dispute, which occurred in Vega Baja, on the island's industrialized northern coast, involved the development of a recreational fishing club and its facilities in an area where commercial fishers had historically stored and launched their vessels. The second, which occurred in Rudy's home of La Parguera, on the Southwest Coast, involved an attempt to create a marine sanctuary in an area where a gradual process of unregulated gentrification had begun to pollute productive nursery grounds and inland waters that fishers had utilized for many years.

## First Dispute: The Vega Baja Yacht Club Versus Commercial Fishers

In the attempts of organized recreational fishers to expand their claims over marine resources—particularly their rights to highly prized game and food fish, such as marlin, grouper, and snapper—they have, for decades, tried to restrict commercial fishing and commercial fishers in the Caribbean and throughout the United States. Often this occurs as tourism develops in the coastal zone and leisure capital begins to structure more and more coastal activity. Conflicts between recreational and commercial fishers in Puerto Rico are varied; yet the most disruptive to commercial fishing livelihoods are those that stem from changing access to marine resources. Privatization of large stretches of coastline—through such developments as marina and resort construction—often lead to access difficulties for commercial fishers. The Vega Baja case serves as an example of a conflict over access in Puerto Rico and the type of mitigation procedures in which the state becomes involved.

### The Setting and Background of the Conflict

The Club Náutico de Vega Baja, the marine recreational fishing club involved in the dispute described here, is located along the North Coast, in a region that includes Arecibo, one of Puerto Rico's most populated municipalities (see Valdés Pizzini, Gutiérrez Sánchez, and Chaparro 1988). This region, composed of ten municipalities, has marinas and yacht clubs only in Arecibo. Recreational fishing infrastructure, such as launching ramps, is limited (almost nonexistent) in the other municipalities. At the time of the dispute, there was a need for marinas and ramps that could offer safe launching and protection of boats. Most commercial facilities and recreational clubs were located in riparian areas of the various estuaries, offering protected spaces for vessels in the absence of natural coves and havens.

Historically, commercial fishing in this area depended on small-scale operations for line trolling in the open sea, the use of beach seines and gill nets, and the use of fish weirs in the estuaries. Local fishers still use most of these, except for fish weirs, which were prohibited in 1953. Field visits to the area's *villas pesqueras* revealed that most were unused or the facilities needed to be upgraded. Commercial landings in the region have been low compared with the rest of the island. In Vega Baja, there was a *villa pesquera* in the area of Cerro Gordo, and Puerto Nuevo Beach had a large ramp for fishers and boaters alike. Except for problems of maneuverability, the ramp was in good enough condition to be of use to the local people.

### The Club Náutico de Vega Baja

The Club Náutico de Vega Baja is one of several nautical clubs around Puerto Rico that provide various kinds of facilities and activities for sportfishers. Like *villas pesqueras*, nautical clubs vary considerably in terms of their membership levels, launching and storage facilities, social activities, and ability to sponsor recreational fishing tournaments. Some of the most successful nautical clubs host international sportfishing tournaments and provide bases for fleets of charter boats and yachts. Smaller clubs may consist of little more than a wooden structure near a natural launching ramp.

The Club Náutico de Vega Baja is located close to the Cibuco River. In May 1987, it boasted 110 members, who paid a membership fee of one hundred dollars and a monthly fee of five dollars. At that time, the club had only a small clubhouse with a bar and a ramp; it lacked piers, slips, and dry-storage areas. Only a large unpaved parking space served the members who used the ramp. The clubhouse and facilities were built by the club members. The ramp, which was built at a cost of eight thousand dollars in the early 1980s, was for the use of the club members, but it was also open to commercial fishers when available. In fact, when the club director was asked about use of the facilities, he responded that the facilities were open for the use of both commercial and recreational fishers. Yet at that time, the conflict between both sectors was at its peak.

The club membership included people from all social classes, who described themselves as "professionals, recreational fishermen, municipal employees, retirees, and commercial fishermen." According to a letter written by the director to the governor, the members were representatives of "all the community's social and economic classes. . . . We have [members] from janitors to factory managers. . . . The membership . . . is not, nor it has ever been, an exclusive elite of the privileged, but a heterogeneous group from our society." This is important because class antagonism has always been a key variable in fishery conflict, including within the recreational sector (Valdés Pizzini 1987).

Because of its structure and membership composition, the club had limited economic resources and depended on monthly dues and annual tournaments for its survival. As in all such clubs in Puerto Rico, social functions and activities carry an enormous weight in the historical existence of the organization. Throughout the year, the club sponsored fishing tournaments for the children (surf fishing) and wives of its membership, tournaments for blue marlin and dolphin, and an island Blue Marlin Classic tournament. Although not profitable, this last tournament was successful in social terms, drawing a large number of visitors and participants.

The club is located in a recreational fishing desert, but its proximity to several beaches and hotel resorts that attract large numbers of tourists led not only to a rise in the real estate values of coastal land in the area around the time of the dispute but also to difficulties between boaters and swimmers. Club members, aware of this fact, pushed for the construction of a marina as the club's home base. A project of this magnitude, however, would require an investment of three to twelve million dollars, an amount not available to the membership at the time of the dispute. They could not have raised this much capital without forming a partnership with a developer, which is, in fact, the route they decided to take. They secured an option to buy the land contiguous to the clubhouse, which connects with the river mouth. Unfortunately, however, the plot of land sought by the club would, at that time, be difficult, if not impossible, to buy. According to various government officials, the land was part of a large farm whose inheritance was the source of a legal dispute among members of a family referred to as the Rodríguez Heir Group.

### History of the Development of the Club

The Club Náutico de Vega Baja started in 1975 with nine members, and its membership increased to eighty in just two years.

Members used the area of the Puerto Nuevo Beach until the volume of boats and boating began to pose safety problems for swimmers. At this time, the club invested money in the ramp, "which," they claimed, "was not public." (Members who recall club history tend to justify this and other behavior on the grounds that the club greatly benefited the community by repairing and maintaining facilities and by infusing the community with capital.) After the difficulties between swimmers and boaters reached a critical mass, club members decided to find another, more secluded, area for club operations.

The ideal location for the club was along the shore contiguous to the mouth of the Cibuco River—the very area utilized for several years by small-scale commercial fishers—in a plot that the government had appropriated from Andrés Rodríguez. Rodríguez, a member of the Rodríguez Heir Group, negotiated the sale of the land to the club. He had a summer home on the plot, in a beach area secluded from the public, but a municipal road ended on his property. Once the land was appropriated, the area was abandoned, serving (as part of it still does) as little more than a junkyard, which made it an ideal location for criminal activities. Joe Martin and Bill Woods, sportfishing commentators for the *San Juan Star* described the saga of the club as follows:

> What was once a junkyard for stolen cars and a somewhat lurid love lovers lane is now a clean, well-maintained clubhouse. The members put in long hours and as much as they could save from $5 monthly dues and paved the road, built water and power lines, repaired the fences and built a boat ramp. Rather than hide these facilities for select few, they have always kept them open to non-members who are able to behave themselves. (Quoted in Valdés Pizzini 1989)

In a letter to the governor, club members emphasized that they made the proper moves and arrangements to purchase the land where their clubhouse was located and the land where they wished to settle. Between the lines, almost imperceptible to the reader, was the fact (as ascertained by officials of CODREMAR and by other informants) that they were occupying that land without any permits from the concerned government agencies. In other words, they were illegal squatters with no legal rights, who were allowed to remain only by the good graces of the government agencies involved. Squatting is not uncommon in coastal or other regions throughout Puerto Rico, as will become particularly clear in the case that follows. Here, however, the fact that the club did not have legal title to the land took on more and more importance as commercial fishers began to make claims with respect to the use of the facilities.

To understand the club's position, it is necessary to be fully aware that the process of privatization and commoditization of coastal lands is well established in Puerto Rico (Valdés Pizzini, Gutiérrez Sánchez, and Chaparro 1988). In his Strategic Plan for Puerto Rico Marine Sportfisheries, Erasto Nieves of the DNR, explains that privatization has "resulted in the exclusion of the public from using the coastal zone for recreational purposes" (quoted in Valdés Pizzini, Gutiérrez Sánchez, and Chaparro 1988:4). The long history of coastal privatization lies beyond the scope of this case; however, in social and political terms, access and privatization are key issues in the political arena and were of central importance in the Vega Baja dispute.

In their plea, club members argued that their investment in club facilities improved the conditions of the area. They claimed that many abandoned structures were rebuilt and occupied by their owners after electricity and water became available. Although club members took much of the credit for bringing electricity and water to the area, these were actually provided by the relevant government agencies and constitute neither evidence of permits nor the implication of land ownership. Government officials often provide public services to squatters to ease conflicts, to provide adequate housing opportunities, or to repay political and economic favors.

Once the area had been improved, club members appealed to the town's mayor to help them obtain the land. When this failed, club members attributed it to a "central government controlled by the opposite [to the major] political party" and to the "fierce monster of government bureaucracy upon our shoulders." In the meantime, improvements to the infrastructure continued, with the club constructing a ramp. The ramp eased the transit of boats along the Puerto Nuevo Beach and provided a needed facility to recreational fishers and club members, who had access to the ramp for a fee. Local commercial fishers could use the ramp without charge.

On June 16, 1987, the club was informed that CODREMAR intended to build a ramp "for a small group of people," (that is, the commercial fishers) in the same lot where the club planned to build its marina—the same lot, that is, under purchase negotiation between the club and the Rodríguez Heir Group. Since the construction of the ramp was to take some time, CODREMAR's Director Sánchez, requested the assistance of the club to allow the commercial fishers to use their ramp.

The club's reply to Sánchez's request underscores both the willingness of club members to work with commercial fishers and the problems of ineffective communication between the two groups. Commercial fishers would not use the ramp because of its long distance from the fishers' community; its insecure, isolated location; and problems associated with using an alien property. Nevertheless, the offer was kept for those "bona fide fishers" (who had licenses and provided landing records to CODREMAR), under the condition of "good behavior" for those who used the facilities, bylaws that also applied to club members. These conditions, of course, implied access under terms agreeable to both the state and the club leadership, implying that those who engaged in "bad" behavior (however defined) would be denied access to club facilities.

Ten days later, the town's mayor entered the dispute, asking the director of CODREMAR to stave off any actual and potential conflict that the impasse was likely to cause. The mayor, in line with the position of the central government, also requested that Sánchez reconsider the site for the commercial fishers' ramp. He did not believe that "CODREMAR's intention was to abort such an important project," since the club had promised to turn the ramp over to CODREMAR once the marina was finished. Throughout these proceedings, most of the promises, intentions, plans, and commitments were without legal support. Neither commercial fishers nor the club had title to the land where the marina was proposed. Thus, along with such issues as the nature of political discourse, this case revealed the pressing need for recreational fishing infrastructure in the region and the problems that the development of such infrastructure would generate. Here a full-fledged conflict was based on the use of one ramp and the potential development of facilities that had not yet been committed to an architect's drawing board!

The meeting between the club's board of directors and CODREMAR's executive director ended on a sour note, because of the "arrogant display of power" of the CODREMAR director, who had his mind set on building the ramp and did not allow the members of the club to explain their grievances. The members of the club, waging war against CODREMAR, wielded their political influence and provided the governor of Puerto Rico with a full account of their side of the case. Nearly twenty days later, advisors to the governor were already working on the problem. In the interim, for reasons unrelated to the case, the secretary for the DNR requested the CODREMAR director's letter of resignation. Months later, shortly after his appointment, the new CODREMAR director met with club members to hear the case.

As a result of a property search that determined that the plot of land on which the club was located belonged to the state, making the club's occupation of the site illegal, the position of CODREMAR changed slightly. The adjoining plot, where CODREMAR had proposed construction of the fishers' ramp, belonged to the Rodríguez Heir Group and its value surpassed the amount that both CODREMAR and the municipality had raised for its purchase and for the construction of the facility. In a meeting with the members of the club, the new CODREMAR director explained that, at the request of six fishers, his agency was considering removing the club from the premises and opening the facilities for public use—specifically, for use by commercial fishers. In a letter, the club's attorney pleaded for reconsideration of the issue based on the economic importance of the "150 club members deserving . . . a real government support" over the welfare of a "reduced group of doubtful classification as commercial fishers." In political terms, the issue was the weight of 150 families of all socioeconomic classes against six families of fishers. In those terms, the decision was difficult indeed.

## Conflict Mitigation

The precarious position of the club, and its alleged good faith and good will to negotiate, forced a conflict mitigation meeting with all sides involved. CODREMAR arbitrated. Club members and commercial fishers were both represented by attorneys. The agenda for the meeting was to go over some of the points that they had agreed upon verbally at a previous meeting. Most important, the club was to provide to fishers, by all possible means, access to the ramp. The legal document presented by the commercial fishers was to serve as a template for discussion. It listed several demands, including that the commercial fishers who belonged to the *villa pesquera* of Vega Baja have access to the ramp and to the parking area and keys to a gate and that the use of the facilities be regulated by representatives from both groups in a democratic fashion.

After a thorough discussion of a few lines of text and some phrasing, both parties agreed with most of the points raised in the document. One area of disagreement involved commercial fishers' use of the facilities. The club did not want to restrict use of the ramp to *villa pesquera* members; it wanted to extend its use to all commercial fishers in Vega Baja. The fishers with the association declared that they did not want to be associated with other fishers, explaining that they did not wish to be legally responsible for the actions and usage patterns of unaffiliated peers. This revealed the bitter conflicts between commercial fishers in Vega Baja, a predicament in which many associations and informal groups find themselves in Puerto Rico. These conflicts can occur for many and varied reasons, but they often derive from each group's relationship with the state (in this case, CODREMAR). The club pressed its position, however, and the final document incorporated a clause stating the availability of the ramp to all Vega Baja commercial fishers, allowing the club to avoid being caught in the middle of what members perceived as a petty feud between local fishers.

Mitigation of the conflict began on the right track. Procedurally, the sides worked together to forge a template for action, with both parties agreeing on a meeting and discussing the issues peacefully. Both attorneys negotiated in good faith and concurred on all issues, even agreeing to establish a mitigation committee to solve future problems involving interpretation of the document and use of the facilities. Although picture perfect in a procedural sense, developments behind the scenes endangered the negotiations.

It was always uncertain how the problems that faced Vega Baja's recreational and commercial fishing community would evolve after the dispute. Casting a shadow over every decision was the reality that the DNR had a de facto moratorium on marina construction. First, although the Army Corps of Engineers was willing to provide permits to those who complied with all regulations, the DNR was reluctant to allow the number of marinas to increase. Second, any permit for a marina in the Cibuco estuary went against a plan for the construction of a canal and possible rerouting of the Cibuco River. Third, the Rodríguez Heir Group was in an unending legal dispute that inhibited any land transaction. These three factors worked against the plans of the Club Náutico de Vega Baja to expand its operations, have a place of its own (with property rights), and be separated from commercial fishers or other resource users. The club was in a catch-22: It could not leave; yet its stay was insecure, particularly without the necessary capital for marina development.

There were also unexplained motives behind the commercial fishers' actions. They could have used other facilities closer to home, such as the *villa pesquera* in the Puerto Nuevo Beach; yet they wanted the facilities in the Cibuco River, in the same space desired by the club. Throughout the mitigation meeting, the commercial fishers' spokesperson assumed an arrogant, defiant attitude toward club members. If it was true that the club was illegally occupying the lot, these fishers had no more right to occupy the area than did the club. The president of the association commented that the issue was to be solved soon by the federal authorities; he planned to call them to remove the club from the premises. The goal was to have the club abandon the facilities and the association occupy them.

*Summary*

The Vega Baja case illustrates a Puerto Rican manifestation of a common dilemma among commercial fishers worldwide: disputes that surround coastal access in areas that are experiencing the multiple and varied processes of coastal gentrification against a background of internal disputes among commercial fishers. Often, recreational fishers occupy the helm of coastal gentrification, privatizing the shoreline or marine resources by restricting access to the sea or, through fisheries regulation, restricting access to the resources that the sea provides.

Yet leisure uses of the coast—resort development, whale watching, coastal golf course development, and so forth—restrict commercial fishing in a variety of ways. Although the Club Náutico de Vega Baja represented one avenue, the dispute in La Parguera illustrates another way in which threats to commercial fishing manifest themselves. It is a case that is entangled in competition among sanctuaries—one a sanctuary for fish, the other a sanctuary for seasonal residents—each chipping away at the refuge that fishers find in fishing.

## SECOND DISPUTE: LA PARGUERA

### Background

La Parguera, a rural barrio, sits in the municipality of Lajas, on Puerto Rico's Southwest Coast. With a resident population of approximately two thousand people in a four-square-mile area, La Parguera is not only an important fishing center; it is also a highly active and growing center for tourism in Puerto Rico's western region. Although La Parguera served as a minor port for the Spanish Crown during the eighteenth century (Cardona Bonet 1985), the original settlement was founded during the early nineteenth century. Early settlers exploited mangroves and reefs, supplementing fishing by cultivating maize and raising cattle, horses, and goats (Feliu 1983:263). *Parguereños* also worked in nearby sugarcane fields and salt ponds. Although these means of livelihood still exist in La Parguera and neighboring fishing villages, most *parguereños* are now employed in commerce, industry, tourism, government, and service jobs outside the community.

Most long-time residents, the settlers of La Parguera, have long traditions of working in the community, but recent developments have brought in several new, seasonal residents. Most important for our discussion, over the past couple of decades, La Parguera's lovely coastline has experienced the growth of houses constructed illegally in the mangrove forests and out over the waters along the shore. These stilt-house *casetas* were built and are occupied primarily by professionals, most of whom have permanent residences outside La Parguera, in larger cities such as Mayagüez or even as far away as San Juan. Thus, the *casetas* are mostly weekend beach houses, with only a few occupied throughout the year. Small boardwalks and piers connect some to the main dirt road. Others are reached by boat. One of the most pressing problems with these structures is that they lack sewage facilities. All human feces end up in the bay, increasing pathogens. The main residences in La Parguera, by contrast, are firmly built on land, most in government *parcelas,* and have cesspools.

Like other coastal communities in Puerto Rico, in the U.S. Virgin Islands, and in the United States (Valdés Pizzini, Gutiérrez Sánchez, and Chaparro 1988), La Parguera has experienced a critical problem of gentrification, which displaces local inhabitants and dominates access to the shoreline with boathouses and *casetas.* This situation increases local pressure on resources and limits physical and visual access to the sea. In essence, coastal gentrification is part of the wider problem of commoditization of the coastal zone, alienating access and privatizing public lands, to the detriment of traditional, coastal fishing settlements (Valdés Pizzini, Gutiérrez Sánchez, and Chaparro 1988).

Today the economic mainstay of La Parguera is tourism. The community's harbor and lodging facilities, the beauty of its sea and landscape, and its public beach attract thousands of tourists all year around. One of the main tourist attractions is a nearby bay whose bioluminescence is produced by a large population of a dinoflagellate (*Piridinium bahamenses*) that emits light when the water is agitated. Several boats in the community offer nightly trips to the bay area, and along with these brief voyages, particularly on weekends, La Parguera's many restaurants, bars, and other attractions bring throngs of visitors and traffic into the streets of the town. At its height on a weekend night, La Parguera comes alive with crowds as thick as a theme park's on opening day, greeting visitors with neon lights, salsa music, and aromas of dishes ranging from seafood pastries to pizza. Tourism in La Parguera has stimulated the development of souvenir shops, marine equipment stores, guest houses, bars, dive shops, guide services, hotels, and seafood restaurants. Out of ten regions that we examined in the NMFS study, Lajas and its neighboring towns ranked first in facilities such as ramps, piers, and waterfront areas; second in the number of marinas and fishing and yacht clubs; and third in the number of services and businesses that serve marine recreation in general (Valdés Pizzini, Gutiérrez Sánchez, and Chaparro 1988).

La Parguera's fishers use primarily traps, diving/spearfishing, and gill nets, exploiting the multispecies environments of the coastal shelf, including mangrove forests, reefs, grass beds, and rocky bottoms of the coastal waters. Most gear is manually operated, although many fishers haul traps with mechanical winches. Traps have been their primary gear, but in the past decade or so, diving for lobster, conch, and reef fish and using gill nets have increased in La Parguera and throughout the island. As is common across Puerto Rico (see Chapter 3), most of the fishers' vessels are eighteen to twenty-one feet long and are propelled by outboard motors. Fishing outings take place daily, leaving early in the morning and returning around noon. Fishers sell their catch to the local *villa pesquera*, to local fish vendors, and to restaurants. Because of the tourist trade, restaurants usually offer the highest prices. La Parguera's marine resources are also used by recreational fishers and divers (spearfishing) and by artisanal fishers from the adjacent settlements of Papayo and Hornos and from the municipalities of Guanica to the east and Cabo Rojo to the west.

## The Critical Event

In January 1983, as we have noted, the DNR and the Sanctuary Division of the NOAA proposed that an area of 68.27 square nautical miles off the Southwest Coast of Puerto Rico be converted into a marine sanctuary. The DNR/NOAA objectives were to protect the marine environment, change the status quo of resource conservation on the island, and provide recreational facilities for visitors yet keep the environment open to some commercial fishing. Fishing remained an essential part of activities allowed in the marine sanctuary, and the sanctuary could indeed "enhance the long-term ability of the resources to sustain the local fishing industry" (NOAA 1983:8). However, the proposed changes would, at least in the short term, adversely affect fishing: Fishers' access to certain areas in the mangrove forest would be restricted by regulations related to its preservation. Proposed underwater trails in the reefs might also constrict fishing territories. Plans to research queen conch and the effects of fecal contamination, coral extraction, spearfishing, and other human activities on the reef created fear that such investigations would lead to the further closing of fishing areas.

Because of perceived ambiguities, commercial fishers of La Parguera opposed the creation of the sanctuary. When the NOAA and the DNR announced public hearings to assess the document, La Parguera fishers began to meet with the people in the community and with fishers from the nearby settlements. They also hired legal and scientific consultants in order to develop their strategy of opposition.

### The Political Environment

Puerto Rico has a high level of electoral participation. The three principal political parties represent different ideologies that relate largely to the question of the island's status vis-à-vis the United States (commonwealth, statehood, or independence; see Chapter 3). Three other indicators suggest that Puerto Rico is a highly politicized society. First, for most government projects and for political and socioeconomic issues, public hearings are held and the people's opinions are taken into consideration in the final decision. Second, citizens lobby for their interests at every level of government. Third, political issues are well covered by the media, which is committed to complete coverage of issues that affect the political balance of power.

Because Puerto Rican political parties continually try to lure away one another's constituents, politicians examine every political cause that may attract their actual or potential constituency. Since the balance of power between the two most powerful parties, the NPP and the PDP, is not clearly defined, the potential for each to increase its constituency is radically affected by current political and social issues. In La Parguera, fishers faced the government apparatus using the democratic mechanisms provided by the political structure in order to protect commercial fishing and their tradition of open access to the coastal environment. These actions were structured through two associations.

## Villas Pesqueras *and Political Praxis*

In our discussion of *villas pesqueras* in Chapter 3, we noted that effective fishing association leadership depends on a president's standing in the community, speaking abilities, skill at balancing political interests, and willingness to maintain the appearance of political neutrality on many issues. These characteristics surfaced in important yet distinct ways during the conflict over the sanctuary. The organized opposition of Puerto Rican fishers to the sanctuary follows from a history of struggle against limited entry (the common practice of limiting entry into a fishery through licensing regulations and other mechanisms). Here two associations represented local fishers' interests. These associations were headed by presidents who conformed, to a great extent, to the "ideal type" described in Chapter 3. One of the leaders had labor and migration experience in the United States and possessed the necessary rhetorical skills, in both English and Spanish, to be an exponent of fishers' problems. (Two years after the dispute, he was elected the fishers' representative for the Caribbean Fisheries Management Council.) Both presidents remained politically neutral throughout (and after) the affair; yet they skillfully used political organizations in defense of their collective interests. In community meetings, elaborating and delivering their messages, both presidents demonstrated different skills with dissimilar styles. The younger of the two assumed an aggressive stance, reassuring everyone that he was willing and able to struggle to the end. He delivered his message in both English and Spanish, making it clear that he fully understood the official documents. The older of the two assumed the position of an elderly man, unskilled in letters and understanding, who needed a thorough explanation of the documents but was firm in his resolve to protect the area for use by fishers. Instead of appearing contradictory and detrimental to their cause, the styles complemented one another, creating a no-win situation for government officials who were dealing with these forms of discourse.

The cultural and political contexts were also crucial. It was an election year, the media offered an outlet for the fishers' grievances, and both the incumbent politicians and their opponents were willing to listen to their complaints. Although peculiar to Puerto Rico, these events parallel other cases of fishers who have fought against the reduction of their traditional economic activities by the state, particularly those that result from the formation of parks and reserves. Such cases have been fought in the U.S. Virgin Islands and North Carolina, among other places (Griffith 1999; Johnston 1987; Koester 1985; McGoodwin 1990).

Key to understanding fishers' arguments is the critical value of class-oriented political discourse. One important issue to emerge in the meetings and activities was the fishers' character as independent producers, similar to laborers or a rural proletariat, in contrast with the middle-class backgrounds of the resource managers and the squatters who occupied the mangrove forests. The fragile character of fishers' economic endeavors forced them to protect their fishing areas from government intervention. As with the rogue association in Pozuelo (see Chapters 4 and 5), many class issues raised in the sanctuary dispute derived from fishers designation of themselves as members of a working class, vis-à-vis the larger society and the state, a designation that provided key semantic elements for their political discourse. Class arguments, real or created, are essential in political praxis and discourse, and the invocation of class affiliation among fishers is a quite conscious ploy to locate themselves in an advantageous position in the social and political arena.

## The Politics of Fishing: Discourse and Strategy

As soon as officials announced public hearings, the two fishing associations from La Parguera called for a joint prehearing meeting to exchange ideas with other fishers. The meeting was held at the town's small one-story community center. The associations contracted the services of the Agencia de Servicios Legales, a government agency that provides free legal assistance. Agency personnel were at the meeting to discuss the legal aspects of the project and to give advice to fishers concerning proper legal procedures and strategies to be taken in dealing with the NOAA and its representatives in the Commonwealth of Puerto Rico. Also present were community members, the mayor of Lajas, small-scale and commercial fishers from the Southwest Coast, and the representatives of a powerful association and confederacy of fishers from Aguadilla, on the northwestern coast.

The presidents of the associations cosponsored and hosted the activity. They opened the meeting by presenting the problem and introducing the people invited to the activity. Both presidents emphasized their united opposition to the marine sanctuary unless its creation would encourage, rather than restrict, commercial fishing. This generated a discussion of the possible effects of the sanctuary. A brief look at the substance of this discussion reveals how the issue of preserving marine resources was transformed into a political struggle over access and entry to the fishery.

Fishers from La Parguera demonstrated highly developed rhetorical capabilities in presenting their position concerning the project. Their arguments, essentially political and economic, were also related to the DNR management of the project. One primary concern was fishers' lack of precise knowledge about what a marine sanctuary entailed. In their view, a marine sanctuary was a protected area, "sacred and virgin," in which economic penetration was not allowed. One of the presidents commented that he had visited a marine sanctuary in Florida and observed that the habitat was well protected but fishing in the area was prohibited.

According to the fishers, the members and representatives of the NOAA and the DNR had met briefly with them in the past to introduce the idea of a marine sanctuary, but the concept and its possible consequences for fishing were never satisfactorily explained. In one meeting, the discussion was held in English with only a rough Spanish translation of the proceedings provided by some university and government personnel at the meeting.

The fishers' principal concerns related to the possible government appropriation of "their" habitat and resources, and the subsequent collapse of their fishery. They resented the fact that the government saw no need to consult with them properly. Most admitted answering a questionnaire that dealt with environmental preservation and the need for preserving the fishery with a management plan. Though the questionnaire stated that it came from the Sanctuary Division, the fishers felt ignorant of the magnitude and intricacies of the project. The fishers also resented the fact that their attitudes were considered for the opinion poll but they were not consulted as the management plan was being developed. In their view, "fishers are the people best suited" for the development of an effective resource management strategy and should have therefore been consulted. Unfortunately, at that time, resource managers were fully prepared neither to discuss the social, economic, and political costs and benefits of the sanctuary to the community nor to allow public participation in planning or an accurate social impact assessment. There is a growing literature that documents the importance of those procedures in third world development (Derman and Whiteford 1985; García-Zamor 1985) in a wide variety of rural and urban settings in the industrialized world (compare Finsterbusch and Bender-Motz 1980; Wulff and Fiske 1987); and certainly for the tropical fisheries (Munro and Smith 1983; Pollnac and Sutinen 1979; Stoffle 1986). At the same time, the importance of local, traditional environmental knowledge of those who rely on natural resources to survive has recently been recognized in policy circles, research agendas, and other settings (see the special 1999 *American Anthropologist* dedicated to the work of Roy Rappaport). This constitutes a clear departure from earlier policy studies that derived from modernization theory and its obsolete economic underpinnings.

La Parguera fishers elaborated several interesting forms of political discourse. For example, they expressed their dissatisfaction with the U.S. government's intervention in "their" environment by alluding to the hot political issue of sovereignty. Another argument presented the project as transnational capitalist conspiracy. An expression heard during the meeting was "the capital is moving around," implying that a profitable private venture—one that desired the breakup of their fishery—was behind the sanctuary. In the fishers' own view, the sanctuary would jeopardize their economic position and their fragile dependence on marine resources, which would be yet "another blow to the Puerto Rican poor working classes." In this way these petty-commodity producers usurped the logic and semantic essence of the radical, pro-independence parties' political discourse. Despite its apparent uniqueness to Puerto Rican politics, fishers' representation of themselves as working class, poor, and powerless vis-à-vis capital is common among fishers around world (Durrenberger and King 2000; Griffith 1999; Griffith and Dyer 1996; McGoodwin 1990; Smith and Jepson 1993).

*The Structure of a Political Event: Public Hearings Strategy*

According to the fishers' strategy, opposition to the marine sanctuary had to be presented as a unified effort to pressure the government, via the public and the media, to legislate in their favor. Firm opposition from every fisher and community member was necessary. To accomplish this, fishers from La Parguera needed the support of fishers from nearby municipalities, most of whom supported their cause. When the confederacy of the Northwest Coast also joined the opposition to the marine sanctuary, an informal fishers' alliance was formed.

The first public hearing was held in the community center. Structurally, it resembled public hearings held throughout the United States. Community members, fishers from all over the West Coast, politicians, government officials, and members of the media attended. Outside, fishers formed a picket line made up mostly of friends and relatives as fishers waited to speak at the hearing. Inside, from a table in front of the main hall, the secretary of the DNR directed the proceedings. She was assisted by personnel from the DNR and from the NOAA. Armed guards from the DNR's Rangers flanked the table and closely monitored the order during the activity. After an introduction to the goals of the hearing and its rules, the secretary listened to the opinions of more than sixty people.

First, La Parguera residents were united in opposition to the sanctuary, arguing that it could destroy fishing and tourism. Two members of the community, both *caseta* dwellers (lawyers with private practices in a nearby town) were also opposed to the project and even questioned the validity of the proceedings. Their main complaint was that development of the sanctuary required an agreement between the U.S. Army Corp of Engineers and the Commonwealth of Puerto Rico. As a result, the *casetas* would become public property, and further construction of the stilt houses would be prohibited. The exponents defended the interests of the absentee and resident *caseta* owners, whom fishers considered responsible for the contamination of the seawater. This conflict of interests between the owners of the *casetas* and the fishers may be the reason that, in the alliance presented by the fishers and the residents, the *caseta* owners did not appear as part of the unitary block, despite their opposition to the sanctuary.

## Fishers as Working Class

At the meeting, forty fishers presented opposing opinions to the committee, focusing on six main themes: (1) The plan was aimed at the destruction of the poor working classes; (2) the marine sanctuary could enable the displacement of a working class (the fishers) by middle-class professionals (the sanctuary managers); (3) it could destroy a three-hundred-year-old cultural and economic tradition in La Parguera; (4) the project favored tourism over fishing; (5) it did not protect the interests of the fishers; and (6) the establishment of the marine sanctuary could end in the total alienation of resources by institutions and agencies of the United States, raising the ever-present issue of Puerto Rican sovereignty to the forefront of La Parguera consciousness.

Again, situating themselves in the working class was central to the fishers' interaction with the state (Kearney 1996; McCay 1984). Fishers stressed their claim to fishery resources as part of their heritage. Emphasizing working-class affiliations, they portrayed themselves as the most legitimate users of the resources, not as recreational users or tourists, creating a dichotomy between production (or work) and leisure. In Puerto Rico, public opinion is more inclined to support those who fish for income, especially when their rights are viewed as attached to the use rights of Puerto Rican peasants: The fishers are an occupational group revered by the media for embodying essential elements of traditional Puerto Rican culture, perhaps the last of the true *jíbaros* (peasants, rural folk). Last, by invoking the possibility of U.S. intervention, the fishers pinched the most sensitive nerve of Puerto Rican politics: questions of sovereignty and Puerto Rico's status—always a field of bitter controversy and unsettled disputes.

## Fishers' Politics, the Media, and the Political Arena

The hearings triggered massive media coverage. Articles, reports, letters to the editor, and interviews with DNR personnel constantly appeared in support of or in opposition to the marine sanctuary. Inevitably, political parties became entangled in the dispute. The NPP (pro-statehood party) maintained its endorsement of the marine sanctuary and the DNR, both essential pieces of its government program. The PDP (pro-commonwealth party) kept a prudent distance and silence for a short period of time, until the pro-independence parties withdrew, at which time it offered to help the fishers. The PIP, with the most political ground to gain, became directly involved when the fishers from La Parguera requested the party's legal and professional assistance. According to one PIP member, the fishers stated that the party was the only political organization able to defend their interests, even though the political proclivities of the fishers were quite different from those of the pro-independence party. Members of the PIP who counseled the fishers' associations were the two top representatives of its elite, both with legal educations and experience in international affairs and forums. For the PIP, it was another occasion to represent community groups in a situation when it appeared that the sovereignty of Puerto Rico was being jeopardized by U.S. intervention.

One of the leaders of the PIP wrote a column in a leading island newspaper, defending the fishers' position and criticizing DNR management strategy. Two weeks later, the secretary of the DNR wrote a reply in the same column, providing a short history of the management programs for La Parguera, created and developed by members of all parties and ideologies since 1917. According to the secretary, the project for the marine sanctuary would provide the proper legal and bureaucratic mechanisms that might finally enable the establishment of a conscientious management plan for the protection of the marine resources and of the fisheries in general.

During that period, the governor of the island visited the town of Lajas for the opening ceremonies of a new public school. The governor also wanted to meet with the fishers in order to receive their direct feedback, and a meeting was arranged. However, the meeting was canceled because fishers objected to the presence of the secretary of the DNR. In an interview with the press, the governor condemned the attitudes of the fishers, commenting that the fishers were being manipulated by persons alien to the fishery. This was an obvious reference to the help they received from the PIP. The governor also reaffirmed his support of the secretary and the DNR.

Various members of the local political elite of the NPP in the municipality of Lajas expressed opposition to the sanctuary because, first, the project could endanger the socioeconomic welfare of the community and, second, it could endanger their political aspirations. For them, a possible solution was to persuade the governor to withdraw his support of the DNR and of the sanctuary, but they were unsuccessful in that effort. Although they were opposed to the project, they did not state their position openly in the press or to the public. Their loyalty to the party was far greater. This is understandable in view of the fact that they were an incipient local political elite that was not yet established solidly in the party's structure. On an individual basis, however, they expressed their feelings to the presidents of the associations. One aspect of the process that this elite could not overcome was the fishers' alliance with leftist groups, an alliance that might prevent the NPP from supporting them in the future.

## CONCLUSION

The disputes in La Parguera and Vega Baja illustrate how fishers drew heavily on established methods of interaction with the state and other, more powerful interest groups; used settings such as public hearings and media-invoking strategies; and represented themselves as low-income, working-class individuals for their own purposes. These overt and covert political party affiliations show the fishers to be highly politically astute and capable of marshaling political capital to further their own interests. Garnering informal support of local leaders behind the scenes has been a particularly effective method of neutralizing opposition in the late twentieth and early twenty-first centuries, one that meshes well with manipulating mass media in ways that draw on issues dear to the public's heart. Issues such as local history and heritage make particularly rich grist for these mills.

In the process of subverting while manipulating these established modes of discourse and confrontation, Puerto Rican fishers legitimate them. The question thus arises whether Puerto Rican fishers' behaviors constitute forms of resistance or enable the processes of class differentiation that might radically alter small-scale commercial fishing in Puerto Rico. In most parts of the world, including the coastal United States, public hearings and associated political processes have not prevented coastal gentrification. Indeed, often these political procedures facilitate gentrification by creating images of democratic participation and fairness that allow displacement and replacement to occur without disturbing the social peace. By lending legitimacy to the very legal and political processes that are likely to undo them, as coastal property continues to attract the attention of ever more powerful people, Puerto Rican fishers may be ultimately undermining the material conditions for their reproduction.

Yet the situation is not that straightforward. Placing political process in the service of domestic production succeeds in giving local political and economic actors some power to assign value to entire ways of life. Instead of merely legitimizing the political process, they may be channeling it toward the support of fishing and, more important, fishing heritage. This is a delicate balancing act, but it is one that may find support, if perhaps for perverse reasons, from processes that we witness around the world at the beginning of the twenty-first century: the increasing importance of cultural brokerage, multiple livelihoods, and the results of partially deterritorialized existences that enable people to marshal support in a variety of settings (as is apparent in the Vieques case in Chapter 8). These behaviors imply constant movement among domestic producer operations and wage-work settings that are more temporary, less secure, and ever more dependent on subsidies from domestic producers for the reproduction of labor. Ironically, by resisting local gentrification, Puerto Rican fishers may in fact be supporting an established, institutionalized, and invasive political economy on the global stage. Yet at home, they insist that their heritage is a critical component of the value of the Puerto Rican coast.

From La Parguera in the Southwest, to the island municipality of Vieques to the east, fishers and traditional coastal settlers remain fundamental custodians of coastal environs. It is ironic that, because of that historical and cultural role, they also present obstacles for development and, for many—including many policy makers—they are a barrier to progress. Writing on communities of the Chesapeake Bay and the Albemarle and Pamlico Sounds of North Carolina, Griffith (1999) argues that fishers and coastal settlers are custodians of a lengthy history of exchanges and reciprocity with nature based on a set of rules much different from those guided by profit and the market. These rules and behaviors achieve livelihoods closer to what some might call sustainability. This history also contains a rich knowledge of nature, which is critical to its understanding and to the appreciation of and conservation of coastal landscapes and marine seascapes.

Similarly, fishers and coastal settlers in Puerto Rico are the stewards of not only a vast knowledge of habitats, species, and ecological dynamics but also an appreciation and historical perspective of the changes in nature (Giusti Cordero 1994). Fishers also feature behaviors of territoriality and spatial utilization that reveal age-old conservation practices and ethics and a sense of the true commons with the active participation of community members in defining sea tenure and usage patterns (Jean-Baptiste 1996; Posada et al. 1996). More important, historically, they have been responsible for leading disputes over coastal issues in favor of the protection of the environment, and they will continue to play that role.

La Parguera's confrontation with the commonwealth and the federal government provided local organizations with the strength to contend their rights to have clean coastal waters. It also established the artisanal fishers of the West Coast as an important political force, committed to maintaining negotiations with commonwealth agencies and disputing decisions they perceive as threatening for the protection of coastal waters. The local fishers, jointly with scientists, supported several actions toward the establishment of a Marine Fishery Reserve, with the understanding that action was needed on their behalf for the conservation of the stocks (Martínez and Valdés Pizzini 1996). Joining the fishers from La Parguera, fishers from Boquerón, San Juan, Fajardo, Vieques, and Salinas have also formed organizations to protect their areas from the assault of leisure activities and infrastructure. Illegal use and occupation of the coastal zone by middle- and upper-class "squatters" has been a problem for the fishers in other areas, such as Guánica, Papayo in Lajas, Ponce, and Las Mareas in Salinas—all communities nestled into the realm of the quiet, beautiful, and productive fringe mangrove forests of the southern coast. Against the social forces of leisure, with varying rates of success, local residents, communities, and environmental organizations have taken a vertical stand to stop the illegal use of the coastal zone and protect access to the coast.

# 8    Fragments of a Refuge

IN APRIL 1999, residents of the island municipality of Vieques, Puerto Rico, drew international attention when, in protest, they colonized a portion of the U.S. Naval Base that takes up about half of the island. For nearly sixty years before the protest, the Navy had routinely used the waters surrounding Vieques for target practice and had allowed other nations access to its firing ranges. Island residents have always considered the shelling environmentally destructive, destroying and disrupting life around the reefs, and on several occasions the bombs have ruined fishing gear.

The April colonization of Navy lands, this time stimulated by the death of a civilian guard, marked the latest in a series of formal and informal protests against the naval presence in Vieques. On each occasion, protest raised the issue of Puerto Rican sovereignty, inevitably invoking particularly sensitive issues of territory and cultural identity among Puerto Ricans dispersed over several regions of the United States, the Caribbean, and elsewhere (Duany 2000). Military power has permeated Puerto Rican history for most of the twentieth century and beyond, from General Davis's 1899 report (quoted in Chapter 2), to the granting of U.S. citizenship to Puerto Ricans in 1917 (just in time for their inscription into the armed forces to serve in World War I), to the U.S. Navy's dominant role in the early decades of U.S. rule. On Vieques (as discussed in Chapter 6), the military's expropriation of land occurred at a particularly disruptive period in the island's history. After immigration to the island for employment in sugarcane first created a demand for food crops and fisheries products, the navy's exercise of eminent domain coincided with the demise of the sugar industry and widespread unemployment. This left the Vieques people with little choice but to help the navy build its base.

These experiences, which left many Puerto Ricans with a negative attitude toward the military, were exacerbated by the disproportionate numbers of deaths of Puerto Ricans in Korea and Vietnam. Furthermore, the issue of Puerto Rico's political status—and the implications of that status for autonomy and sovereignty—is the one that most strongly determines party affiliation and voting and occupies the forefront of Puerto Ricans' political consciousness. Indeed, nonvoting Puerto Rican members of the U.S. Congress, commenting on the standoff between the residents of Vieques and the navy in the days before the residents' arrests in May 2000, agreed that the confrontation never would have taken place had the island had either state or independent status. As the situation stood, the actions of the Vieques people against the U.S. Navy received wide approval and support across the island and throughout the Puerto Rican diaspora, cutting across class and political party lines. The zenith of this support occurred when, as part of her bid for a seat in the New York senate, former first lady Hillary Clinton visited Vieques to back those who were colonizing navy lands, directly opposing her husband by suggesting that the navy cease all shelling and target practice.

Although this may have been the most recent and most visible assertion of locals' rights to the lands and waters of Vieques, the residents of Vieques engaged in several similar struggles against the navy and other powerful forces during the twentieth century. As suggested by the fishers life histories presented in previous chapters, fishing families often spearheaded struggles, in Vieques and elsewhere across Puerto Rico, using resources at the household and community levels to organize resistance to processes that threatened their access to launching sites and coastal and marine resources. The continued conflicts between Vieques fishers, other coastal residents, and the U.S. Navy highlight the myriad problems that small-scale fishers face along today's rapidly changing coastlines.

Chapter 7 pointed to the problems that occur in coastal regions as they experience the immigration of relatively wealthy individuals and groups who promote economic apartheid by privatizing parts of the coast and reducing access to resources traditionally used by local commercial fishing families. This process, usually a by-product of conventional tourism, has occurred throughout the Gulf and South Atlantic states of the United

States, in parts of Mexico, and in other places around the Caribbean and Central and South America (Griffith 1999, 2000; Smith and Jepson 1993). Yet the April 1999 problems in Vieques, particularly when considered against the background of those who were dislocated from work in sugar, help illustrate the diversity of contemporary difficulties that face commercial fishers and other native coastal dwellers and the diversity of their responses to these difficulties. Coastlines are not only generally pleasant, desirable places for tourists; they are also strategic border regions of ports and harbors that, increasingly, transport the huge volume of goods bought and sold electronically. Heavy commodities such as wood chips, petroleum, sugar, and grain depend on shipping. Dealers of illicit drugs and smugglers of all kinds, in turn, depend on coasts and the diversions of shipping, as well as on coastal real estate to launder money. In what was, only a decade or so ago, a quaint seaside Puerto Rican village called Boquerón, a construction boom is underway today. Fueled in part with illegal activities, it has laid the foundation for the growth in gated communities and condominiums that are certain to attract retirees and seasonal residents—more people, that is, like the *caseta* owners of La Parguera. These residents, in turn, will most likely demand a range of health, recreational, and other services that may complement or conflict with the livelihoods of Puerto Rico's commercial fishers at the same time that they will increase the island's current demand for immigrants from the Dominican Republic, Haiti, and other Caribbean locations. We witness similar changes shaping the physical and social landscapes of coastal regions across the Americas. Again and again, they raise this question: Where do coasts and coastal dwellers, and especially commercial fishers, lie in the so-called "new" economy of high-tech transfers; the expansion of leisure capital; our continued dependence on ports and shipping; and the accompanying changes that smugglers, seasonal residents, retirees, and new immigrants demand?

Earlier we pointed out that Puerto Rican fishers are better positioned than most of us to critique, with their ways of life, the transition from political and religious oppression and consolation to the impassioned and directionless lashing out of dislocated, alienated people that Paz characterized as the world's spiritual wilderness. Puerto Rican fishers, like Puerto Ricans generally, are among the original transnationals, identifying with a widely dispersed diaspora. At the same time, Puerto Ricans are among the last colonial subjects, daily awash in questions of sovereignty, autonomy, and self-rule. As nation-states disintegrate around us, as Paz's "buried passions of tribes, sects, and churches" (1995:263) threaten national allegiance everywhere, Puerto Ricans find themselves divided between the pioneering social forms of transnationalism and the antiquated political relationships of colonialism (Duany 2000). At once empowered and overpowered, free and constrained, they enjoy relatively more opportunity than others in the Caribbean and Latin America but still suffer from the precarious whims of a dominating political economy, a most powerful nation-state.

This condition and the ambivalence it generates, Puerto Rican historians remind us, is not entirely new.

In a 1986 book entitled *Los gallos peleados* (*The Fighting Cocks*), Puerto Rican historian Fernando Picó argues that, from 1910 to 1940, in the Puerto Rican municipality of Utuado, legislative initiatives succeeded in outlawing several domestic producer enterprises. Home milk manufacture, home cigar and cigarette manufacture, home brewing of liquor, cock fighting, running games of chance, native folk curing, and street vending all became illegal or highly constrained. Official moves against these enterprises were justified on the grounds of either public health or child welfare. Because home-produced milk, cigarettes, cigars, liquor, and remedies, as well as the commercial traffic in goods sold on the street, could not be adequately inspected or monitored, officials claimed that they posed a potential health threat to consumers. At the same time, many home-based enterprises used child labor and at times even actively drew neighborhood children into their production regimes.

Both the use of child labor and the production and sale of goods and services that pose potential threats to public health are common features of small-scale fishing in Puerto Rico. Indeed, these are common features of many small-scale-producer activities based in households everywhere. Because seafood has been known to cause mild to severe illness and, occasionally, death, fishers are particularly vulnerable to accusations that their products could jeopardize public health. Furthermore, in the Caribbean, the neurotoxin ciguatera, which builds up in some species of fish (such as trunkfish) that feed around coral reefs high on the food chain, is a constant potential health hazard.

Yet it is not primarily on the basis of threats to public health that Puerto Rican fishing, as a domestic producer enterprise, is being challenged. Rather, it is on the basis of threats to environmental health. Like many commercial fishers around the world, Puerto Rican fishers today have been losing ground to urban growth, real estate development, and other dimensions of the commoditization of the coastal zone—a process that has been justified, in part, by the portrayal of commercial fishing as a threat to the health of the marine environment. Routinely, fishers are accused of depleting fish stocks and, as in the case of Pozuelo, of damaging mangroves and other nursery grounds to gain access to the sea, build traps, and set and land nets (Griffith and Maiolo 1989).

In his consideration of domestic producer activities, Picó traced the ironic emergence of several new, "dysfunctional" social groups to the rubble of criminalization; he calls these groups, quite appropriately, *los perdedores*: the losers. He goes on to list and discuss several groups in this category: alcoholics, drifters, homeless people, prostitutes, mentally disturbed individuals, and others for whom the formal labor force offers few opportunities. Evidently, in Picó's case, improving public health involved sacrificing mental health.

Given our observations regarding the therapeutic aspects of fishing, particularly as a remedy for injuries of the formal economy, the criminalization of fishing in Puerto Rico will most likely contribute to problems that Picó found in Utuado in the last quarter of the twentieth century. These problems, of course, remain part of today's social landscape, in Puerto Rico and elsewhere. As ironic as are the trade-offs between public and mental health that Picó discusses, the forces that confront small-scale fishing today are no less contradictory: Developments in the global labor market encourage the maintenance of small-scale fishing as a refuge from and nursery for capitalist labor markets, on one hand, and developments in local Puerto Rican coastal environments encourage the fragmentation of fishing communities into diverse groups that bicker over smaller pieces of a shrinking pie, on the other hand.

## Labor Markets, Diasporas, and Puerto Rican Fishers in the Newest International Division of Labor

We noted in Chapter 1 that, in the 1970s and 1980s, social observers began to write about a new international division of labor based on the hypermobility of capital and labor. Multinational corporations, able to shop around for the lowest-cost and most-productive workers, had harnessed shipping and communications technologies to manufacture products in several countries. As capital moved, so did people—across regions and international boundaries in search of work. They flooded labor markets in developed regions and nations at the same time that highly paid, unionized workers suffered plant closings and reduced job security from the flight of capital.

While these developments took advantage of new technologies and new organizational forms, there was little new in the search for low-cost and productive labor or in the practice of recruiting workers from distant locations to undermine the bargaining positions of empowered workers. Indebting workers though company stores or consumer loan programs; encouraging households to subsidize the reproduction of labor with small-scale farming and fishing; achieving higher levels of labor control through company housing and transportation policies; enforcing labor policy compliance through violence and intimidation; conscripting child labor; segmenting workforces by ethnicity, race, gender, or bureaucracy; and encouraging workers to enlist household members to achieve dictated productivity levels—all of these practices, and more, have been features of the social landscape of capitalist labor relations since the early days of capitalism. The faces may have changed, and some of the specifics, but the guiding principles of power remain.

In social science jargon, *flexible accumulation* refers to capitalist activity that incorporates workers' households unevenly across space and time. Its defining feature is its ability to expand or contract production based on market, labor, and regulatory considerations. The flexibility to engage and disengage labor is a cornerstone of this activity; yet it depends on reducing alternative economic avenues in such a way that potential workers continue to need outside employment at the same time that they continue to maintain cultural refuges capable of sustaining the unemployed. Flexible accumulation also depends on the ability of employers to establish short-term, specific relationships with workers, instead of encouraging lifetime commitments in return for job security and the feelings of *communitas* that derive from being a part of the same organization throughout life.

Such precise contracts are relatively easy to negotiate for low-skill, low-wage tasks, although these contracts are becoming more common among some professional workers as well. Those of us in universities are well aware of the growth of instructor positions that take advantage of the large numbers of underemployed people with doctorates. These positions offer one- to three-year contracts to fresh scholars to teach a course here, a course there, as enrollments rise and fall or as tenured faculty press for reduced teaching loads in order to pursue research and writing. Many high-tech corporations, such as IBM, have laid off employees only to rehire them as independent contractors. These arrangements are often seen as highly desirable to both employees and employers. From the employee's perspective, the arrangement allows him or her to initiate similar independent contractual arrangements with a variety of other companies and puts the employee more in control of his or her own time. From the employer's perspective, the arrangement allows the company to reduce costs associated with health insurance, pensions, unemployment insurance, office space, computing equipment and supplies, and other direct costs of maintaining employees.

Although it is as difficult today to speak of a totally new international division of labor as it was in the 1970s or 1980s, these contractual arrangements are signs that the trends that social observers noted in the last quarter of the twentieth century are expanding across the global economy and becoming institutionalized. Training programs for new workers poised to enter the labor force have recently come to adopt several new buzzwords that reflect the realities, or perceived realities, of employment. The phrase "a lifetime of learning," for example, as opposed to "learning for a lifetime," underscores the point that individuals need to continue to learn throughout their lives, because it will be necessary to retrain and retool our human capital skills to respond swiftly to shifts in capital investment. Students are advised to "keep options open" and always "seek new opportunities," phrasing that puts a positive spin on a negative social development: For most, the days of a secure career with a single employer or a single trade are over.

Among highly skilled workers, the loss of job security may be counterbalanced by the high prices that highly skilled consultants are able to charge for their services, at least at the height of their productivity, as well as by whatever sense of independence they derive from shopping around for the highest bidder. These qualities tend to be unavailable to workers without opportunities to acquire highly paid skills or impressive credentials. Workers who operate from weaker positions vis-à-vis established norms of human capital formation, as we have seen again and again in previous chapters, engage in activities that draw upon several family members and depend on informal ties with other households. Where small-scale subsistence and income-generating opportunities exist, these tend to be sifted into the full body of household and community livelihoods—both to achieve accepted levels of subsistence security and, in the case of fishing, to relieve tensions that derive from working at the low end of the wage-labor market.

For people from these backgrounds, at this moment in history, subcontracting agreements are less like the agreements struck between IBM and its former employees (which may approximate friendly partnerships) than like those agreements forged in work settings with which many Puerto Ricans have been all too familiar. Exemplary work settings include perishable crop agriculture and construction. In these settings, work groups tend to be small, flexible, hastily organized and equally hastily disbanded, often based on kinship and friendship networks, and increasingly drawing on workers from several regions around the world (Griffith et al. 1995).

Most Puerto Rican fishers, accustomed to such work settings, never enjoyed lifetime job security. Puerto Rican fishers have always combined small-scale production with several other kinds of economic pursuits, formal and informal, legal and illegal, low paying and grueling, or relatively easy and completely safe. Coming from backgrounds where seasonal economic pursuits dominated the economy until recently, Puerto Rican fishers seem almost preadapted to labor markets in which employers encourage workers to vanish during periods of economic contraction. Puerto Ricans' heritage of migration and transnationalism, elaborated over the past century, further enhances this ability to move around inside economic formations as opportunities open and close. Although this may make it seem as though Puerto Rican fishers were particularly well poised to leap successfully into the twenty-first century, remember that many of these adaptations developed under a variety of largely negative conditions and power imbalances. If the life histories presented throughout this volume leave us with nothing else, they should teach us that Puerto Rican fishers engage in multiple livelihoods and traverse geographical space for many reasons that have less to do with individual desire than with their positions in households and fishing communities whose contours have been heavily influenced by North American capitalism. In many of their experiences, specific sets of tasks, tied to a specific industry or set of industries, have become deeply embedded in wider, ongoing processes of survival that include fishing as they engage more and more members of the fishing community in these other, nonfishing industries and tasks. In this day and age, especially among Puerto Ricans, these communities are likely to be dispersed across several regions or nations. They are likely to become part of, in a word, a diaspora—its members transnational in their experience and influence (as Hillary Clinton's efforts in Vieques showed), linked to one another as much by senses of a shared past, identity, and cultural expression as by the kinds of human relationships that help people select lovers and companions, break bread, and make their livings.

The development of transnational collectives or communities has received a great deal of attention in the social sciences recently, in part because deterritorialized social structures—whether households, factories, or communities—have become common features of late-twentieth-century and early-twenty-first-century social landscapes (Basch, Glick-Schiller, Szanton Blanc 1994; Portes 1997). Again, transnational communities, like many of the features of the so-called new international division of labor, are not wholly new phenomena. Since early in the twentieth century, Puerto Ricans in particular have been members of communities very much like transnational communities, even when they did not cross international boundaries (Duany 2000; Morris 1995). Other groups, such as Mexicans and Mexican Americans along the California and Texas borders or Portugese fishers in New Bedford, Massachusetts, have enjoyed the ease of movement and active participation in communities at home and abroad for about a century as well (Foner 1997).

With the interest in transnationalism, observers of immigrants and international labor migrants have begun to turn away from distinctions between home and abroad, paying closer attention to the ways that migrants and refugees keep in close contact with their homelands, creating fluid social structures that seem to transcend territory. In conjunction with several studies of immigrant-dominated labor markets and industries across the United States, for example, Bach and Brill (1990) found that employers had only to place calls to Guatemala, working through the networks of a group of immigrant Guatemalans based in Houston, Texas, to fill jobs in Houston supermarkets within seventy-two hours. In their influential work, Basch, Glick-Schiller, and Szanton Blanc (1994) discuss the experiences of Haitians, Filipinos, Vincentians, and Grenadians in terms of their historical propensity to live fully in more than one location. Whether they actually live in more than one location or only as part of an imagined community, these transnationals nevertheless have constructed, in opposition to and as part of hegemonic processes, "deterritorialized nation-states" (Basch, Glick-Schiller, and Szanton Blanc 1994). Politicians from countries with highly dispersed populations (that is, diasporas) have acknowledged the importance of their expatriot communities by traveling to overseas locations to solicit their support.

As transnationalism has become more widespread with the proliferation of forms of communication, forms of travel, and legal mechanisms for crossing borders, it has enhanced the diffusion of Spanish-speaking peoples throughout the United States. In rural areas, the dramatic growth of Latino/a populations has been particularly notable (Naipaul 1989). Most of these immigrants arrive from Mexico, but many come from El Salvador, the Dominican Republic, Nicaragua, Cuba, and Guatemala. Recently, new indigenous Mexican and Guatemalan sending regions, particularly in Oaxaca and Chiapas, have emerged to supply agriculture, food processing, landscaping, construction, and other rural industries with increasingly ethnically diverse workforces. Many of these groups include political refugees, whose ability to communicate with relatives and friends at home is greatly dependent on political developments in their homelands and on U.S. foreign policy. In these situations, relationships between the people who remain behind and the people who flee may take on mythical characteristics, as cultural practices and beliefs that have diminished in importance at home remain powerful in ethnic communities abroad (Foner 1997). This situation exists even as these population movements inspire new consumption practices and undermine native paths to authority and respect.

In the United States, Puerto Ricans take potentially strategic positions in relation to the growth of transnational communities based in the Caribbean and Latin America. They are often skilled in English and Spanish and occupy neighborhoods and labor markets that new immigrant Latinos/as are entering. The semiproletarianized condition of many Puerto Rican fishers' households, combined with their members' linguistic skills and their transnational experience with fluid, long-distance social structures, again, may preadapt them to moving between domestic production and wage labor during this most recent phase of global capitalism. Does this imply, too, that Puerto Ricans are relying less and less on the United States? Might they be well situated to take advantage of a more nebulous social formation—one noted for its involvement with several branches of an informal economy that reaches throughout Latin America, encouraging bilingualism among its members along with a shared sense of a creolized Spanish, African, and Native American past?

Perhaps. Yet several scholars have pointed to the fact that, in plebiscite after plebiscite, Puerto Ricans reject independence from the United States although they continue to consider themselves quite distinct, particularly culturally, from the vast majority of U.S. citizens. These sentiments have been elaborated and promoted to levels of political, intellectual, and popular discourse that reify Puerto Rican cultural nationalism, hispanophilia (the love of things of supposed Spanish origin), sovereignty struggles, and divided identities. Yet much of this discourse leaves behind what daily life for Puerto Rican working families has taught North American capitalism. These so-called essentialist conversations among elite Puerto Ricans miss the mark: Much of the production of culture occurs in various kinds of economic and ecological bedding, within class struggles such as those that we witness in La Parguera yet also in relationships that link households with the marine environment.

Throughout this text, we have used the metaphor of the journey to depict, describe, and interpret the life histories, stories, and labor trajectories of Puerto Rican fishers. This metaphor plays with the ambiguity born from both the general and particular details of their lives: Are they fishers at work or workers at sea? In the microhistory of their lives, we came across the richness and depth of their agency and the myriad processes and positions that they occupy in the social structure. Yet their lives suggest a more or less continuous resistance to structure and a longing for pathways different from those that other wageworkers have taken. Despite this situation, however, their lives are similar to the trajectories of laborers throughout the world, and their actions are inextricably linked to the workings of capitalism and the position that Puerto Rico occupies in the global context and in relation to the United States.

By these observations, we confront the problem of identity: the "true" character of these people's lives. We set out to study the lives and labor experiences of Puerto Rican fishers. More than twenty years of field experience, in Puerto Rico and elsewhere, led us to use extreme caution in our representation of fishers and in our interpretation of the varied and subtle ways that they are defined by anthropologists, sociologists, government officials, and themselves (Griffith 1999:xi–xii). Clearly, the trap of containment—expressed by Michael Kearney (1996) as he reconsidered peasants—was set. In our social scientific training lurks the need to contain, define, encapsulate, and thus freeze fluid and open social relationships and process into categories and types that we can manage, describe, and submit to statistical analyses in order to create nomothetic models or forms of explanation. Working under such constraints, fishers are (not surprisingly) those who fish and practice their livelihood in relation to the coastal and marine ecosystems, extracting and exploiting marine resources through a wide variety of schedules, methods, and market regimes. Yet they are also something else. We believe that fishers, similar to others in the anthropological literature, are fluid people who dodge the clear-cut categories that petrify their constant movement; chameleonic lifestyle; and polysemous, variable life histories. Latin American peasants (Kearney 1996), Colombian indegenous people (Taussig 1987), and the !Kung of the Kalahari (Gordon 1981) provide a handful of ethnographic examples of highly mobile peoples, interpreted in a variety of ways, never fully contained, but always subjected to a policy of containment on behalf of social scientists, archaeologists, the military, rubber barons, farmers, and travelers. Never fully understood, they remain a blur. What ties them together—among many other characteristics—is their capacity for physical mobility, inserting themselves in complex and dense landscapes, moving in and out of the panoptic eye of the state. Nomads, swidden tribes and agriculturalists, peasants in the tropical forest using slash and burn, maroons, gatherers and hunters, and fishers weave livelihoods that occur alongside of, at times as parts of, dominant production forms and markets. The landscape is their accomplice; it forms an inextricable relationship between these groups and nature and spaces, a relationship that reaches well beyond the obviously utilitarian. More than spaces, these landscapes are converted into *places*, with deep cultural meanings, sometimes becoming the topography that maps identity (Harvey 1993; Maldonado, Valdés Pizzini, and Latoni 2000). In this sense, the landscape (or seascape) becomes a platform where a specific identity is formed and maintained, despite the multiplicity of forms in which people traverse other spaces, modes of production, and types of labor. And in this sense, forest and wilderness are homes to those who prefer to live at the borders. Those who inhabit the forest remain either unfathomed or imagined as wildmen, or they are simply labeled as the enemy of progress and capital.

Juan Giusti Cordero's (1994) dense analysis of the ambiguous category of peasants/proletarians in the landscape of northern Puerto Rico's Piñones, in the heart of the wetlands and mangrove forests once surrounded by sugar plantations, provides an excellent example of this richness of meanings. Giusti Cordero elegantly disputes Mintz's (1956) claim that after 1898 there was a generalized process of proletarianization of coastal peasants into the plantation system structured by U.S. capital. He demonstrates the intense, conscious, and contentious process by which peasants wove a matrix of economic strategies and relationships with the environment to dispute the absolute control of the *colonos* (sugarcane growers), the plantation owners, and the dominant force of the central sugar mills (see Giusti Cordero 1994 for a thorough historical analysis of sugar capital in the Spanish Caribbean). Totally erasing the distinction between microhistory and the *longue durée* of structural processes, Giusti Cordero embarks on the dialectal analysis of the forms and contents of the resistance of peasants and their incorporation into wage labor. In reading Giusti Cordero's remarkable work, we noticed the difficulty of selecting either the category of peasant or the category of proletarian, in time and space, to identify the *piñoneros*. With the historical transformation of the landscapes and ecosystems used for sugarcane production and peasant activities, the *piñoneros'* relation with local and international markets remained fluid and complex, alternating spaces and relations of production in order to survive *la bruja:*

> Like time, space offered few absolute boundaries in Piñones. The social spaces of the villages and the plantation were each far from homogeneous, but were marked by perplexing relationships of opposition and complementarity in their ecology and labor. Thus, while peasant labor processes are abstractly linked to village social space—viewed as dwellings, garden plots and pastures—in Piñones peasant labor, like real village space, ranged far beyond. It encompassed large coconut groves, lagoons, and an extensive mangrove forest, crossed by canals that the *piñoneros* plied on their flatboats, as well as the seashore. (Giusti Cordero 1994:35)

As sugarcane production came to an end, the *piñoneros* faced the difficulties of market exchanges for the commodities, such as charcoal, pastries, and fish, that they produced and exchanged. *Piñoneros* also faced dramatic environmental changes caused by sand extraction, pollution, beach erosion, exposure to coastal hazards, and contamination that affected their longtime relationship with nature. Piñones has also been threatened by the potential development of large tourist resorts, a threat that has hung as the proverbial sword over their heads during the past two decades or so. According to Giusti Cordero, in the midst of these transformations, fishing and its spaces remained the fundamental platform for their livelihood and identity: "Perhaps fishing (along with agriculture) remained always among the most autonomous of 'peasant' activities in Piñones. This suggests that it was fishing, after all—long an 'odd man out' in accounts of peasant life—that remained among the most 'peasant' activities of all" (1994:35).

*Piñoneros*, coastal people of African ancestry and maroon backgrounds, settled in a rich and complex landscape painted with estuaries, wetlands, cays, mangrove forests, and a highly wave-energized coast (see Giusti Cordero 1994:660–708; Sued Badillo and López Cantos 1986). This coastal landscape provided the sources for their seasonal autonomy and independence from the plantations (Giusti Cordero 1994:764). Being proletarians at times and peasants with a feeble claim to the land at others, the *piñoneros* took the opportunity offered by the coastal ecosystems to survive and construct a cultural tradition based on their African ancestry (see Giusti Cordero and Godreau 1992). Wood-cutting, charcoal making, crab catching, and fishing became essential activities that distinguished the *piñoneros* from urban others even as these endeavors linked them to the rest of the peoples of the coast. The obsession of public officials and social scientists with using increasing wage labor as an indicator of progress or economic advancement also leads to the obliteration of the category of fishers. Either fishers become invisible by their own fiat or they are missed by observers. Giusti Cordero comments on the census takers in Piñones in the early part of the twentieth century, observing that they lacked sensibility, failing to identify the true character of fishing and other coastal activities in the local economy, mainly because fishing's seasonal character tended to obscure its importance (1994:772).

In this book we have depicted, to a large extent, circular and complicated labor trajectories of fishers who traverse different habitats and countries as they become inserted in the workings of capitalism. Yet Giusti Cordero remains focused on *local* journeys in time and space—between seasons and throughout environs—and the forms and relations of production in a complex coastal ecosystem that appears as swamps, wetlands, estuaries, and mangroves. In a metaphorical reference to the hybrid character of mangroves as ecosystems, Giusti Cordero also critically assesses the category of peasants/proletarians as a hybrid concept that may undervalue the rich lives of these highly mobile (physically as well as categorically) coastal peoples (1994:792–93). Deep in the ambiguous nature of peasants/proletarians, a number of people resisted work in sugar by engaging in fishing and other productive activities (1994:893). Tinkers/fishers, charcoal makers/fishers, woodworkers/fishers, cane cutters/fishers—these tempting binary distinctions that appear to be endless—mask the complex lives of these apparent rogues, in hiding, engaging, contesting, denying, fighting, and at some point surrendering to wage labor.

## The Fishing Interstice

Again, our essential interest in this anthropological enterprise was to understand the life histories of Puerto Rican fishers. And again, a professional, and in many ways personal, question remains: Were we engaged in an investigation of fishers as an essential category, or were we betrothed to the description and analysis of the many ways in which fishing provided a meaningful and concrete setting in the lives of these workers? It is difficult to say. Nevertheless, one fact remains clear to us: For the Puerto Ricans we interviewed, visited, and shared with, fishing is the key interstice in which they locate themselves in their complex trajectories. This work of ours is not merely the end result of an anthropological study of fishers. Instead, it is an ethnography of the specific pathways where we, armed with the instruments of anthropological research, encountered the fishers in light of our main goal of studying them. In retrospect, we realize that we were not interviewing fishers; we were interviewing people who at different times in their lives found refuge in the interstices of the coastal environs and in fishing as a vocation. For the majority of these people, this was, too, their fundamental and essential cultural context. Fishing is where they originally belong in the social structure, in the *longue durée* of maritime communities. But it is also where they draw upon, manipulate, and defy state and capital. It is the place where their lives are meaningful and the platform from which they embark on a wide and long journey into wage labor. That is what we read from the circular journeys narrated to us.

One characteristic of these stories remains constant: their circularity. Though these journeys of fishers take a long time, they share the common trait of a *return* to the island. In many cases these returns are recurrent; the fishers may return many times from their travels into the fields of New Jersey; the factories of New York; or the refineries in St. Croix, in the U.S. Virgin Islands. Back in Puerto Rico, they are transformed into laborers, fishers, public servants, and perhaps migrants who again fly to New York only to return once more. Recently, after the works of various sociologists (for example, Juan Hernández Cruz [1985]), Jorge Duany (2000) characterized this process as the *vaivén*. A *vaivén* is a pattern of continuous travel back and forth in time and space that clearly maintains memory and identity tied to the original landscape of one's childhood or tied to those meaningful landscapes (or chronotopes)—such as Puerto Rico's tropical rainforest and national park, El Yunque—that bring home the relationship with the nation (see Maldonado, Valdés Pizzini, and Latoni 2000).

In essence, this is an ethnography of highly mobile individuals who are situated in a matrix of labor contexts in the world economy throughout their lifetime. At the same time, they firmly hold to the identity of fishers and coastal settlers as their most meaningful (in the ecological, communal, cultural, historical, and political sense) trait. As anthropologists, we employ the trope of the *periplus*, the narration of a long, circular navigation whose main destiny is a return to the point of departure—the *nostos* (the homecoming) narrated in the poems of return, the *nostoi* of the Greeks (Calvino 1991:22; Kundera 2000:11). Homer's *Odyssey* is one such narrative of return (*nostos*) and its suffering (*algos*), embodied in the term "nostalgia," which is a longing for that which is sometimes unrecoverable: memories, spaces, relationships, experiences (Rosaldo 1989). In his novel *Ignorance* (2000), Milán Kundera claims that the homecoming journey is characterized by a blurred memory of the homeland and a crass ignorance of the processes back home: a curse suffered by most migrants. In Homer's *Odyssey*, Ulysses is unaware of a great deal in Ithaca, and only through difficult effort can he picture Penelope's face. That is the price of a prolonged journey. Yet today, with telecommunications, electronic media, air travel, and all other technical supports of globalization, space and time have been restructured and distances largely obliterated, minimizing ignorance and rendering memory and experience far less difficult (Appadurai 1992:4; García Canclini 1999; Harvey 1996). The recurrence of the return, what Duany calls *el vaivén*, provides the setting for the constant insertion of coastal peoples into the refuge of fishing, when therapy is needed, when massive layoffs take place in manufacturing, when seasonal agricultural work ends, or simply when it is time to return to one's Ithaca.

In essence, who is Ulysses? Is he a prince, a journeyman, a beggar, an explorer, a lover, a penitent? It is difficult to say. It depends on which book of *The Odyssey* one reads, and it depends on time and space. His identity hinges on the encounter, though he remains a man from Ithaca and a prince. In this book, this encounter, we captured the precise moment that these coastal people were back in their Ithaca, safe at the shoreline, embedded in their fundamental identity without being paralyzed by it. In many instances, they hastened away from our field notebooks and interview schedules, because they found jobs in the construction sector or simply because *se embarcaron*—they embarked on yet another journey back to the cities, fields, and nurseries of the United States, abandoning their refuge, leaving momentarily or for a long period of time the interstice of fishing.

In reading the lives of fishers as profound cultural dramas woven through the history of capital and global transformations, we recall the allegory of *The Odyssey* narrated in this book through the life history of Liche (see Chapter 1). It was rather hard not to see Ulysses in every face we met, since most were somewhere along a constellation of the long, circular journey, forever returning home, coming up against all the images and shadows that characterize Homer's text.

In more than one way, a fisher named José Enrique Báez reminds us of the life of the Achaean hero. The first time we met, a hot summer night in 1982, José was in a crowded communal center in La Parguera, several miles away from his home. There he was, ready to give the fishers from the Southwest his support in opposing the marine sanctuary. He spoke softly, briefly, but convincingly about the rights of fishers and the need to protect their social class. The concept of social class, unlike for many of José's contemporary colleagues, appeared to be an essential part of his oratory and an essential part of his long experience as a laborer, away from home, at different points along the Puerto Rican diaspora.

The day we met him, José declined to grant us an interview. Yet as we listened to him and others during that remarkable public hearing, we realized the key importance of the use of "class" as a concept and as an aid in the discourse employed in a contestation to the state. From afar, we watched his careful steps in the political arena, moving the fishers from Aguadilla into a series of counterhegemonic activities, solidifying opposition to the marine sanctuary, and encouraging fishers of the West to join forces with an island-wide organization of fishers in Puerto Rico:

El Congreso de Pescadores del Este (The Eastern Fishers' Congress). This organization was born out of the Vieques fishers' struggle against the U.S. Navy.

When we finally were able to interview José, he was an independent producer, owner of his own vessel and fishing gear. Unlike many other fishers from Aguadilla, José owned a relatively large (twenty-four-foot) boat, which he used to navigate the rough waters between Quebradillas and Mayagüez (basically the northwestern portion of the island) and the island of Desecheo. He used a multispecies, multigear strategy, alternating fishing with traps, longlines, and hand lines for tunas. His son fished with him every day, or—when his son was working another job—his son-in-law covered for him. During our research, José's son was enrolled as a lifeguard in a government job-creating program. José taught his son—in the same fashion that his own father had taught him—how to fish. In the best tradition of artisanship (José's grandfather was a shoemaker, a distinction he makes in his narrative with great pride), José knows how to (and usually does) make and fix his own gear. The story of shoemaking is meaningful in that José's grandfather worked as a shoemaker when the sea was too rough to fish. A fisher and a craftsman, his grandfather was an *ambulante:* Without a stationary store, he took to the streets to offer his services in town. That mobility allowed him to return to fishing when he needed to do so.

One critical and significant day, José's father moved to the United States *buscando futuro* (looking for a [better] future). José moved to New York in 1949, at the height of the Puerto Rican migration to the United States, during the critical period of the transition from an agricultural enclave in decline to the new industrial Puerto Rico. But there he was, a quintessential fisherman sitting before us, enchanting us—as Ulysses did with many others—with his stories of *villas pesqueras*, of fishing in the rough waters of the northern coast, and of his commitment to fishing as a true vocation. With the Aguadilla waves pounding at the shoreline, a few meters from the veranda where we sat, José retold the long journey of the *nostos* once again. For more than twenty years he lived with his family in New York and New Jersey, employed as a laborer in various factories. José constantly returned to Puerto Rico, keeping a link, as many others told us they had done, with fishing and with the coastal communities where he was born. His speech showed no signs of linguistic interference or the use of anglicisms. During the interview, it was impossible to guess that he had lived in the United States for such a long time. Back home, he fished to rekindle his love for the sea. In New Jersey, he invited friends and relatives to go fishing. He was a recreational fisher on a party boat, reenacting his true identity, preparing for the final return, so that he could fish as a refuge for the rest of his days. We can only imagine him as a sedentary factory worker in New Jersey, peacefully watching television in his free time but remembering (as many fishers revealed that they had) the fishes and the tonalities of the seascape. At heart, like many Puerto Ricans, he was committed to *el vaivén*, which committed him and his family to the *nostos* to reinsert himself in the interstice of fishing.

If José appears to be a sedentary person, who stayed for long periods of time on the U.S. mainland, returning only briefly for vacation, Shaolín, a fisher who constantly reacted to the speed and direction of capital, may seem to be hypermobile.

In his small, exceptionally comfortable house in Vieques, this fifty-three-year-old man lived with his wife, and at the time that we visited him, he was finding refuge in the interstice of fishing. The couple seemed to run a tight ship: He dived for conch, and she cleaned and sold the fish at home. As the owner of his boat, he remained an independent fisher who preferred to sell his catch with the help of his wife. He despised the fish dealers from Fajardo who visited Vieques daily to buy fish at low cost. Instead, Shaolín sold directly to the restaurants and small hotels of Vieques, and for a time he sold fish and conch in Saint Thomas and Saint Croix. Born into a fishing family, he joined the U.S. Armed Forces as a paratrooper at age sixteen. He became a sergeant, and he was a champion boxer in the 122-pound division. Licensed to drive in Fort Bligh, Texas, he followed the pathways of the transnationals: truck driver in Houston (1959–1960), truck driver (1960–1962) and taxi driver (1962–1966) in New York (with a brief stint as a mechanic in Chicago and Philadelphia from 1963 to 1964), longshoreman in Texas and Mexico (1965), taxi driver in Saint Croix (1966–1976) and occasional diver, and trap fisher and diver in La Esperanza during his first return to Vieques (1976–1979). In 1979, the conflicts between the fishers and the U.S. Navy, which continue to this day, intensified. A political faux pas alienated Shaolín from the fish market: One day, while he was driving his pickup truck in town, he played a tape of the Davilita and Daniel Santos recording *Los Patriotas,* a collection of *independentista* songs. Everyone heard the songs through the loudspeakers. The mayor (who was from the conservative side of the PPD, or pro-commonwealth party) blackballed Shaolín, and he started to have a hard time selling his fish in town.

Shaolín left Vieques and moved to St. Croix to live and fish. Then in 1980 he moved to St. Thomas to live as a fisher. Once again, "restless spirit" (as he described it) pushed him to Tortola in the British Virgin Islands, where he worked as a mechanic and fished. In 1984, he returned to St. Thomas and found employment as an outboard mechanic at Red Hook Marina, where he served the recreational fishing community. Throughout 1986 he lived and worked in St. John as a mechanic and fisher and in St. Croix as a taxi driver and then returned to Vieques. In 1987 Shaolín went back to St. Thomas to fish and to work as a tour driver until his final return to Vieques in 1988, when our paths crossed at last.

It is evident that Shaolín is no Ulysses longing for Penelope. His domestic life with his wives and children reads like what Arjun Appadurai (1992) terms an ethnoscape, made up of Texans, Cruzans, Cha-Chas (the French fishers of St. Thomas), gringos, and Puerto Ricans. His is the trajectory of a complex and hypermobile life, always on the edge, defying the norms and the coherent descriptions of historians and anthropologists alike. Similar to Maqroll El Gaviero (the sailor), the hero of Álvaro Mutis's novels, Shaolín forged birth certificates (to enter the armed forces at a young age), lived (as he described it) "in slavery" (as a child, he was forced to tend hogs for an aunt in Maunabo after his father's death), gambled, cannot recall with clarity all his children and wives, and one day left Vieques with neither compass nor timeframe, only to find himself in the midst of a fascinating story of which he shared with us a mere fragment.

The life histories of José and Shaolín represent complex lives that, although perhaps not entirely typical, illustrate the varied ways that we encountered fishers at some point in their trajectories. At the precise moment of our investigations, coastal people such as José and Shaolín found themselves in the interstice of fishing, momentarily assuming the role of independent producers. But their personal accounts also read as the long journeys of wage laborers taking part in different diasporas. Yet one fact remains clear: Their true identity is that of the fisher.

In the pursuit of an understanding of this identity, its cultural expressions, and its relationship to wealth and power, many who are interested in diasporas have overlooked the fact that the movement among nations and regions, challenging allegiance and sovereignty, is also movement among livelihoods and productive regimes. The people who participate in these livelihoods, however long or briefly, operate under various terms of trade, create and assign value by different paths, allow individuals greater and lesser degrees of power, and possess different qualities and degrees of attachment to natural resources. The life histories presented here show that the conditions that involve fishing families in fluid, diasporic relations of production influence how they perceive and critique production arrangements and also influence their propensity for collective activity. They remain very attached to their home communities and territories and to fishing ways of life. That attachment makes it difficult to suggest, first, that territory is less important among transnational migrants than it once was and, second, that artisanal fishing and peasant agriculture are separate production realms from those of North American or global capitalism.

We find support for both points in that, in the short term, local developments often frustrate global trends. As fishing communities' roles as sanctuaries from the downswings and injuries of work in capitalist labor markets become ever more relevant vis-à-vis global capital and flexible accumulation, the very qualities that make them sanctuaries are being threatened from both without and within. The fact that coastal landscapes are, indeed, contested terrains highlights their territorial importance. Perhaps more significant, the fact that fishers can appeal to representatives of the state and garner widespread popular support in their struggles suggests that many people outside fishing communities perceive the intrinsic, cultural value of fishing lifestyles to coastal landscapes. A summary of this book and its litany of problems that threaten fishing communities reaffirms the contested nature of coasts and the varied forces of wealth, power, and heritage that take an interest in their future.

In our discussion of Puerto Real in Chapter 3, we saw that, within fishing communities, alliances and conflicts around the marketing of fish promote diversity within the community as they alter the face of commercial fishing. This occurs among fishers worldwide. In some cases, marketing developments have made fishers marginal to commercial fishing; in others, they have reduced fishers to wageworkers within fishing enterprises that have become dominated by the marketing sector. In both cases, fishers become less self-reliant; suffer increased income insecurity; become more exposed to the fluctuations of economic processes that develop outside their communities; and are usually forced to alter production schedules, gear types, and other fishing practices. Responding to these changes, some fishers move into marketing themselves, retaining at least partial control of their fates. Still, many fishers, like those in Puerto Real, who fish for fish houses lose much of the power to make decisions about fishing techniques, targeted species, time spent fishing, places fished, and even the use of fishing practices adopted as resource conservation measures.

As we have discussed in many different ways beginning in Chapter 4, these dislocating and reorganizing forces internal to fishing communities join with other forces—whether because of economic developments or for more personal reasons—and cause fishers to engage in a series of varied jobs (*chiripas*), or multiple livelihoods, to make ends meet. Fishers who are partially marginalized from fishing, particularly when marginalization results in less time to devote to fishing, often shift to fishing strategies that are less labor intensive. They may, for example, set gill nets and traps to soak while they work at a part-time job or attend to another domestic producer activity. In this way, as we saw in Chapter 5, they move between the hazardous environments of labor and the therapeutic environments of fishing at various points throughout their lives. Struggling to leave wage labor forever, as did the Victors and Santoses whose stories were presented in Chapter 6, they may draw upon the collective resources of their households and the households of their sons, daughters, and in-laws to establish viable fishing enterprises. In some cases, they manage to accomplish this even in the face of the U.S. Navy's heavy restrictions on fishing or the opposition to commercial fishing that is issuing, more and more, from the leisure capital interests discussed in Chapter 7.

Again, these many obstacles that face commercial fishers and other coastal peoples drive home the point that coastal environments are particular kinds of landscapes. As Marta Maldonado, Manuel Valdés Pizzini, and Alfonso Latoni (2000) point out in their work on El Yunque, these landscapes are Puerto Rico's tropical forest, *places* instead of mere *spaces*. Drawing on the work of geographer David Harvey and others, Maldonado, Valdés Pizzini, and Latoni write, "Space is thought to denote a natural medium, a generic collection of geographic features, a 'tabula rasa' still to be inscribed with the specificities of culture and history. Place, on the other hand, refers to space that is located and understood in its historical contexts. The notion of place reminds us that geographies do not exist in a socio-political vacuum" (2000:82). Coastal landscapes are actual, tangible places that convey particularly important meanings to commercial fishers and others who have lived for generations along the water. Imagining fishers, as well as others who are heavily dependent on natural resources, as members of deterritorialized social formations obscures the importance of coastal landscapes in fishers' sense of well-being and the reproduction of their ways of life. Through family ties and other kinds of social linkages, people sprinkled throughout the Puerto Rican diaspora receive various material, cultural, and other supports and gifts from these land- and seascapes. Thus, diasporas remain linked not only with specific political entities, territories, and areas of cultural production but—most important here—with specific *natural resources*.

The fact that many small-scale productive activities are intimately tied to natural resources—whether marine, agricultural, wild, or other—involves these households and these diasporas in environmental issues. As suggested by the many protests against the navy in Vieques, the criticisms that fishers of La Parguera and Aguadilla levied against the owners of the *casetas*, and the general complaints against the source of marine pollution that we encountered again and again in the field, fishers view their own fates and the fates of marine habitats through the same lens. Although privatization of coastal environments and marine resources will probably result in capital concentration, it is more troubling to fishers that these processes are very likely to cut off their access, and their family members' and children's access, to marine resources. Too much of a reduction of fishing households will make it increasingly difficult to move between fishing and wagework. A critical mass of fishing households must remain in place for others to come and go across the space and time that capital first organizes and then fishers reorganize to suit their needs.

The suggestion here is not only the importance of territory but the importance of a particular kind of territory, with specific qualities that bind it to ways of life that, in turn, participate in the production of culture. Because households have become paramount in fishing ways of life, the position of women vis-à-vis marine resources and informal economic activities is distinct from the position that they typically hold in capitalist labor markets, in state policies, and in the vision of family often promoted by political and religious groups. In Chapter 6, in particular, we saw women as more than merely supportive in a household's movement away from dependence on sugar and wagework; they were instrumental and often even the driving force of this process.

Their relationship to the men of the fishing world cannot be strictly viewed as one of power, class, culture, or love; rather, it is a complex relationship that links and negotiates among markets, natural resources, and opportunities.

It is the same with fishing families and communities of fishers. In much of today's cultural analysis, a kind of consensus is emerging that culture—or what is left of that concept that once lay at the heart of anthropology—is primarily a product of power struggles, a concept that the previous chapters support in various ways. Yet the chapters herein also suggest that culture is a product of relationships—within and between households and between fishers and marine environments—in which power is less important than are various feelings and identities. The sea's tendency to produce conceptions of relief, therapy, and health combines with fishers' working backgrounds to create components of a fishing culture that is learned and shared on the water.

Puerto Rican fishers may experience, daily, the many ambiguities embedded in the overarching contradictions between transnationalism and culturally significant territory. Yet, in their daily excursions into the sea, they manage to suspend their direct involvement in the anxieties and rewards of struggle and feed their heritage on the great and tiny satisfactions of catching fish. Having opened this work with two quotes from Nobel laureates— one pedagogical, the other horrified—we close with a quote from Elizabeth Bishop's lovely poem *Santarem* (1983:185). Here the speaker encounters a watery environment that inspires the kind of peace that Puerto Rican fishers express when they speak of their fishing, their craft:

> That golden evening I really wanted to go no farther;
> more than anything else I wanted to stay awhile
> in that conflux of two great rivers: Tapajós, Amazon,
> grandly, silently flowing, flowing east.
> Suddenly there'd been houses, people, and lots of mongrel
> riverboats skittering back and forth
> under a sky of gorgeous, under-lit clouds,
> with everything guilded, burnished along one side,
> and everything bright, cheerful, casual—or so it looked.
> I liked the place; I liked the idea of the place.
> Two rivers. Hadn't two rivers sprung

from the Garden of Eden? No, that was four,
and they'd diverged. Here only two
and coming together. Even if one were tempted
to literary interpretations
such as: life/death, right/wrong, male/female
—such notions would have resolved, dissolved, straight off
in that watery, dazzling dialectic.

# References

Abgrall, Jean. 1975. "A Cost Production Analysis of Trap and Hand Line Fishing in Puerto Rico." Ph.D. diss., Department of Economics, University of Rhode Island at Provincetown.

Acheson, James. 1987. The Lobster Gangs of Maine. New Hanover, N.H.: University Press of New England.

Algren de Gutiérrez, Edith. 1987. *The Movement Against Teaching English in Schools of Puerto Rico.* New York: University Press of America.

American Anthropologist. 1999. "Contemporary Issues Forum: Ecologies for Tomorrow: Reading Rappaport Today." 101 (1).

Anderson, Benedick. 1983. *Imagined Communities: Reflections on the Origins and Spread of Nationalism.* London: Verso.

Antler, Ellen, and James Faris. 1979. "Adaptation to Changes in Technology and Government Policy: A Newfoundland Example (Cat Harbour)." In *North Atlantic Maritime Cultures,* ed. Raúl Anderson, 129–54. The Hague: Mouton.

Appadurai, Arjun. 1986. "Introduction: Commodities and the Politics of Value." In *The Social Life of Things: Commodities in Cultural Perspective,* ed. Arjun Appadurai, 3–63. Cambridge, England: Cambridge University Press.

——. 1992. *The Condition of Modernity.* Minneapolis: University of Minnesota Press.

Bach, Robert, and Howard Brill. 1990. "The Impacts of the 1986 Immigration Reform and Control Act on the U.S. Labor Market and Economy." Final Report to the U.S. Department of Labor. Binghamton: State University of New York Institute for Multiculturalism and International Labor.

Basch, Linda, Nina Glick-Schiller, and Cristina Szanton Blanc. 1994. *Nations Unbound: Transnational Projects, Postcolonial Predicaments, and Deterritorialized Nation-States.* Langhorne, Pa.: Gordon and Breach.

Beckford, George. 1972. *Persistent Poverty.* New York: Oxford University Press.

Benería, Lourdes, and Gita Sen. 1982. "Accumulation, Reproduction, and Women's Roles in Economic Development." *Signs* 7:279–98.

Bishop, Elizabeth. 1983. *The Collected Poems of Elizabeth Bishop.* New York: Farrar, Strauss, and Giroux.

Blay, Federico. 1972. "A Study of the Relevance of Selected Ecological Factors Related to Water Resources and the Social Organization of Fishing

Villages in Puerto Rico." Mayagüez, Puerto Rico: University of Puerto Rico Water Resources Institute.

Bonilla, Frank, and Ricardo Campos. 1981. "A Wealth of Poor: Puerto Ricans in the New Economic Order." *Daedalus* 110:133–76.

———. 1985. "Evolving Patterns of Puerto Rican Migration." In *The Americas in the New International Division of Labor*, ed. Steven Sanderson. New York: Holmes and Meier.

Bourgois, Philipe. 1989. *Ethnicity at Work*. Baltimore, Md.: Johns Hopkins University Press.

Brandes, Stanley. 1975. *Migration, Kinship, and Community*. New York: Academic Press.

Brush, Stephen. 1978. *Mountain, Field, and Family*. Philadelphia: University of Pennsylvania Press.

Buitrago, Carlos. 1972. *Esperanza*. New York: Wenner Gren Foundation.

Burton, Michael. 1972. "Occupational Beliefs." In *Multidimensional Scaling*. Beverly Hills, Calif.: Sage.

Calavita, Kitty. 1992. *Inside the State*. New York: Routledge.

Calvino, Italo. 1991. *Porqué leer los clásicos*. Barcelona, Spain: Tusquets Editores.

Cardona Bonet, Walter. 1985. *Islotes de Boriquén: Amoná, Abey, Sikeo, Piñas y otros: Notas para su historia*. San Juan, Puerto Rico: Oficina Estatal de Preservación Histórica.

Castells, Steven. 1984. *Here for Good: Western Europe's New Ethnic Minorities*. London: Pluto Press.

Center for Women's Economic Alternatives. 1989. "Annual Report." Ahoskie, N.C.: Center for Women's Economic Alternatives.

Chaney, Elsa, and Marianne Schmink. 1980. "Women and Modernization: Access to Tools." In *Sex and Class in Latin America*, ed. June Nash and Helen Safa, 160–82. New York: J. F. Bergin.

Chavez, Leo. 1989. *Shadowed Lives: Undocumented Mexicans in the United States*. New York: Harcourt Brace.

Chayanov, A. V. 1966. *Theory of Peasant Economy*. Homewood, Ill.: American Economics.

CODREMAR. 1987. *La pesca de Puerto Rico*. San Juan, Puerto Rico: CODREMAR.

Collazo, J. R., and J. A. Calderón. 1988. "La pesca en Puerto Rico, 1979–82." Technical Report. Laboratorio de Investigaciones Pesqueras del Departamento de Recursos Naturales y Ambientales de Puerto Rico 1 (2): 1–30.

Collins, Jane. 1988. *Unseasonal Migrations: The Effects of Rural Labor Scarcity in Peru*. Princeton, N.J.: Princeton University Press.

Collins, Jane, and Martha Giménez, eds. 1990. *Work Without Wages: Domestic Labor and Self-employment Within Capitalism*. Albany: State University of New York Press.

Comaroff, Jean, and John Comaroff. 1992. *Ethnography and the Historical Imagination*. Chicago: University of Chicago Press.

————. 1999. "Occult Economies and the Violence of Abstraction." *American Ethnologist* 26 (2): 279–303.

Comitas, Lambros. 1974. "Occupational Multiplicity in Rural Jamaica." In *Work and Family Life: West Indian Perspectives,* ed. Lambros Comitas and David Lowenthal, 157–73. Garden City, N.Y.: Anchor Books.

*Congressional Record.* 1917. 64th Cong., 2d sess. (January 20): 2248–65.

Cook, Scott. 1982. "Industrialization Before Industrialization." Paper presented at the University of Florida at Gainesville, November.

Dalton, George, ed. 1971. *Economic Development and Social Change: The Modernization of Village Communities.* Garden City, N.Y.: Natural History Press.

Davis, George W. 1899. *Report on the Civil Affairs of Porto Rico.* Washington, D.C.: Department of the Interior.

Deere, Carmen. 1983. "The Allocation of Familial Labor and the Formation of Peasant Household Income in the Peruvian Sierra." In *Women and Poverty in the Third World,* ed. M. C. Buvinic, M. A. Lycette, and W. P. McGreevey, 104–29. Baltimore, Md.: Johns Hopkins University Press.

Deere, Carmen Diana, and Alain de Janvry. 1979. "A Conceptual Framework for the Empirical Analysis of Peasants." *American Journal of Agricultural Economics* 61 (4): 601–11.

de Janvry, Alain. 1983. *The Agrarian Question and Reformism in Latin America.* Baltimore, Md.: Johns Hopkins University Press.

Derman, William, and Scott Whiteford. 1985. *Social Impact Analysis and Development in the Third World.* Boulder, Colo.: Westview Press.

Dibbs, J. L. 1967. "Report of a Marketing Study of Two Puerto Rican Fishing Communities." Bridgetown, Barbados: UNDP/FAO Caribbean Fisheries.

Doeringer, Peter, Philip Moss, and David Terkla. 1986. *The New England Fishing Economy: Jobs, Income, and Kinship.* Amherst: University of Massachusetts Press.

Duany, Jorge. 2000. "Nation on the Move: The Construction of Cultural Identities in Puerto Rico and the Diaspora." *American Ethnologist* 27 (1): 5–30.

Durrenberger, E. Paul. 1992. *It's All Politics.* Mobile: University of Alabama Press.

————. 1995. *Gulf Coast Soundings: Mississippi Shrimpers.* Lawrence: University Press of Kansas.

Durrenberger, Paul, and Thomas King, eds. 2000. *State and Community in Fisheries Management.* Westport, Conn.: Bergin and Garvey.

Durrenberger, Paul, and Gisli Pálsson. 1989. *The Anthropology of Iceland.* Iowa City: University of Iowa Press.

Edwards, Richard. 1979. *Contested Terrain.* Chicago: University of Chicago Press.

Feliu, Cesar. 1983. "Fundación del Poblado de La Parguera." In *Historia de Lajas,* ed. Mario Pagas, 250–72. Lajas, Puerto Rico: Pagan.

Finsterbusch, Kurt, and Anabelle Bender-Motz. 1980. *Social Research for Policy Decisions*. Belmont, Calif.: Wadsworth.

Fitchen, Janet. 1992. "On the Edge of Homelessness: Rural Poverty and Housing Insecurity." *Rural Sociology* 57 (2): 173–93.

Foner, Nancy. "The Immigrant Family: Cultural Legacies and Cultural Changes." *International Migration Review* 31 (4): 961–74.

Forman, Shepard. 1970. *The Raft Fishermen: Tradition and Change in the Brazilian Peasant Economy*. Bloomington: Indiana University Press.

Foster, George. 1969. *Culture and Conquest*. New York: Viking Fund Publications in Anthropology, No. 27.

Frank, Andre G. 1967. "The Development of Underdevelopment." In *Capitalism and Underdevelopment in Latin America*, ed. Andre G. Frank, 67–98. New York: Monthly Review Press.

Friedland, William, and Dorothy Nelkin. 1971. *Migrant: Agricultural Workers in America's Northeast*. New York: Holt, Rinehart, and Winston.

Frobel, Folker, Jürgen Heinrichs, and Otto Kreye. 1980. *The New International Division of Labour: Structural Unemployment in Industrialised Countries and Industrialisation in Developing Countries*. Cambridge, England: Cambridge University Press.

Furtado, Celso. 1976. *Economic Development in Latin America*. Cambridge, England: Cambridge University Press.

Galarza, Ernesto. 1964. *Merchants of Labor*. New York: McGraw-Hill.

García Canclini, Néstor. 1999. *La globalización imaginada*. Buenos Aires, Argentina: Paidós.

García Márquez, Gabriel. 1967. *One Hundred Years of Solitude*. New York: Modern Library.

———. 1992. *Love in the Time of Cholera*. New York: Modern Library.

———. 1995. "The Solitude of Latin America: 1982 Nobel Prize Acceptance Speech." *Georgia Review* 49 (1): 133–36.

García-Zamor, Jean-Claude. 1985. *Public Participation in Development Planning and Management: Cases from Africa and Asia*. Boulder, Colo.: Westview Press.

Gatewood, John, and Bonnie McCay. 1990. "Comparison of Job Satisfaction in Six New Jersey Fisheries: Implications for Management." *Human Organization* 49 (1): 14–25.

Girvan, Norman. 1973. "Caribbean Mineral Economy." In *Caribbean Economy*, ed. George Beckford. Kingston, Jamaica: Institute of Social and Economic Research, University of the West Indies.

Giusti Cordero, Juan. 1994. "Labor, Ecology, and History in a Caribbean Sugar Plantation Region: Piñonez, Puerto Rico, 1770–1950." Ph.D. diss., State University of New York at Binghamton.

Giusti Cordero, Juan A., and Michel J. Godreau. 1992. "Las concesiones de la Corona y propiedad de la tierra en Puerto Rico, siglos XVI–XX: Un estudio jurídico." *Revista Jurídica de la Universidad de Puerto Rico* 62 (3): 351–79.

Gmelch, George. 1987. "Return Migration." *Annual Review of Anthropology* 47:124–56.

Gómez, Ángel Gregoria. 1985. "Consideraciones sobre un modelo de psicoterapia para el puertorriqueño." *Homines Revista de Ciencias Sociales: Tomo Extraordinario* 3:102–10.

Gonzalez, Lugardo Marin. 1985. "La ergoterapia como pilar fundamental en los servicios de salud mental." *Homines Revista de Ciencias Sociales: Tomo Extraordinario* 3:201–25.

Gordon, Edmund. 1981. "Phases of Development and Underdevelopment in a Caribbean Fishing Village: San Pedro, Belize." Ph.D. diss., Department of Anthropology, Stanford University.

Gordon, David, Richard Edwards, and Michael Reich. 1982. *Segmented Work, Divided Workers.* Cambridge, England: Cambridge University Press.

Gramsci, Antonio. 1971. *Selections from the Prison Notebooks.* New York: International.

Griffith, David. 1983. "The Promise of a Country: The Impact of Seasonal U.S. Migration on the Jamaican Peasantry." Ph.D. diss., Department of Anthropology, University of Florida at Gainesville.

———. 1984. "International Labor Migration and Rural Development: Patterns of Expenditure Among Jamaicans Working in the United States." *Standford Journal of International Law* 19 (2): 357–70.

———. 1985. "Women, Remittances, and Reproduction." *American Ethnologist* 12 (4): 676–90.

———. 1986. "Social Organizational Obstacles to Capital Accumulation Among Returning Migrants: The British West Indies Temporary Alien Labor Program." *Human Organization* 46 (1): 34–42.

———. 1987. "Nonmarket Labor Processes in an Advanced Capitalist Economy." *American Anthropologist* 89 (4): 838–52.

———. 1993. *Jones's Minimal: Low-Wage Labor in the United States.* Albany: State University of New York Press.

———. 1995. "Names of Death." *American Anthropologist* 97:453–56.

———. 1997. "Lasting Firsts." *American Anthropologist* 99:23–29.

———. 1999. *The Estuary's Gift: An Atlantic Coast Cultural Biography.* University Park: Pennsylvania State University Press.

———. 2000. "Economic Apartheid and Social Capital Along the Coasts of the Americas." *Urban Anthropology* 29 (3): 255–84.

Griffith, David, and Christopher Dyer. 1996. *An Appraisal of the Social and Cultural Aspects of the Multi-Species Groundfish Fishery in New England and the Mid-Atlantic.* Woods Hole, Mass.: National Marine Fisheries Service, Northeast Regional Office.

Griffith, David, Ed Kissam, Jeronimo Camposeco, Ann Garcia, Max Pfeffer, David Runsten, and Manuel Valdés Pizzini. 1995. *Working Poor: Farmworkers in the United States.* Philadelphia: Temple University Press.

Griffith, David, and John Maiolo. 1989. "Considering the Source: Testimony vs. Data in Conflicts Surrounding Gulf and South Atlantic Trap Fisheries." *City and Society* 3 (1): 74–88

Griffith, David, Manuel Valdés Pizzini, Jeffrey C. Johnson, Ruperto Chaparro, and James Murray. 1988. "A Socioeconomic Analysis of Recreational Fishing in Puerto Rico." National Marine Fisheries Service Technical Report. St. Petersburg, Fla.: National Marine Fisheries Service/National Oceanographic and Atmospheric Organization, Southeast Regional Office.

Gringeri, Christina. 1994. *Getting By: Women Homeworkers and Rural Economic Development.* Lawrence: University Press of Kansas.

Gudeman, Stephen. 1978. *The Demise of a Rural Economy.* London: Routledge and Keagan Paul.

Gudeman, Stephen, and Alberto Rivera. *Conversations in Columbia.* Cambridge, England: Cambridge University Press.

Gunn, Thomas. 1918. *The Maya of British Honduras.* Washington, D.C.: Bureau of American Ethnology.

Gutiérrez Sánchez, Jaime. 1982. "Características personales y de trabajo de los pescadores en Puerto Rico." University of Puerto Rico Sea Grant College Program Report (UPR-SG-85-02). Mayagüez: University of Puerto Rico.

Gutiérrez Sánchez, Jaime, Manuel Valdés Pizzini, and Bonnie McCay. 1986. "La pesca artesanal y las asociaciones de pescadores en Puerto Rico." University of Puerto Rico Sea Grant College Program Report. Mayagüez: University of Puerto Rico.

Hage, Dave, and Paul Klauda. 1989. *No Retreat, No Surrender: Labor's War at Hormel.* Ithaca, N.Y.: Cornell University Press.

Haraldsdottir, Gudrun. 1994. "Women's Work in Industrial Fishing Economies." Master's thesis, Department of Anthropology, University of Iowa at Iowa City.

Harvey, David. 1993. "From Space to Place and Back Again: Reflections on the Conditions of Postmodernity." In *Mapping the Futures: Local Cultures, Global Change,* ed. J. A. Bird, B. C. Curtis, T. L. Putnam, G. J. Robertson, and L. O. Tickner. New York: Routledge.

———. 1996. *Justice, Nature, and the Geography of Difference.* Oxford, England: Blackwell.

Heine, Jorge, and Juan Manuel García-Passalacqua. 1983. *The Puerto Rican Question.* New York: Foreign Policy Association, Headline Series.

Hernández, Raúl. 1985. "Historia de los programas de salud mental en Puerto Rico." *Homines Revista de Ciencias Sociales: Tomo Extraordinario* 3:22–31.

Hernández Cruz, Juan. 1985. "¿Migración de retorno o circulación de obreros boricuas?" *Revista de Ciencias Sociales.* 24 (1–2): 81–112.

Hewitt de Alcantera, Cynthia. 1976. *Modernizing Mexican Agriculture.* Geneva, Switzerland: UNRISD Publications.

History Task Force. 1979. *Labor Migration Under Capitalism: The Puerto Rican Experience.* New York: Monthly Review Press.

Hobsbawm, Eric, and Terrance Ranger. 1983. *The Invention of Tradition.* Cambridge, England: Cambridge University Press.

Holmström, Mark. 1984. *Industry and Inequality.* New York: Cambridge University Press.

Homer. 1974. *The Odyssey.* Trans. Albert Cook. New York: W. W. Norton.

Idyll, Clarence. 1972. *The Potential for Fisheries Development in the Caribbean and Adjacent Seas.* Marine Bulletin 1. Kingston: University of Rhode Island, Center for Marine Resource Development.

Iñigo, Felix. 1968. "El fomento de la industria pesquera en Puerto Rico y sus perspectivas." *Agricultura al Día* 15(3–4): 1–16.

International Labor Organization (ILO). 1983. *Encyclopedia of Occupational Safety.* Vols. 1 and 2. Geneva, Switzerland: ILO Office.

Irizarry, Rafael L. 1985. "El desempleo, la educación y los servicios de orientación a jóvenes: Las contradicciones de un paradigma." *Homines Revista de Ciencias Sociales: Tomo Extraordinario* 3:163–68.

Jarvis, Norman. 1932. *The Fisheries of Porto Rico.* Washington, D.C.: U.S. Department of Commerce, Bureau of Fisheries.

Jean-Baptiste, Neudy. 1999. "Distribución especial de las nasas y sus relaciones con la topografía, agregaciones de peces y capturas estacionales." Master's thesis, Department of Marine Sciences, University of Puerto Rico at Mayagüez.

Johnson, Jeffrey C. 1992. *Selecting Ethnographic Informants.* Beverly Hills, Calif.: Sage.

Johnson, Jeffrey C., and David Griffith. "Visual Stimuli in Research." In *Visual Data,* ed. Elsa Sobo, 72–95. Beverly Hills, Calif.: Sage.

Johnston, Barbara. 1987. "The Political Ecology of Development: Changing Resource Relations and the Impacts of Tourism in St. Thomas, USVI." Ph.D. diss., Department of Anthropology, University of Massachusetts at Amherst.

Kearney, Michael. 1996. *Reconceptualizing the Peasantry.* Boulder, Colo.: Westview Press.

Koester, Stephen. 1985. *Socioeconomic and Cultural Role of Fishing and Shellfishing in the Virgin Islands Biosphere Reserve.* St. Thomas, U.S. Virgin Islands: Island Resource Foundation.

Kopytoff, Igor. 1989. "Singularization and Commoditization" In *The Social Life of Things,* ed. Arjun Appadurai, 1–38. Cambridge, England: Cambridge University Press.

Kottack, Contrad. 1992. *Assault on Paradise.* Boston: McGraw-Hill.

Kundera, Milán. 2000. *La ignorancia.* Barcelona, Spain: Tusquets Editores.

Kurlansky, Mark. 1999. *Cod.* New York: Penguin.

Lamphere, Louise. 1987. *From Working Daughters to Working Mothers.* Ithaca, N.Y.: Cornell University Press.

Leach, Edmund. 1964. *Political Systems of Highland Burma.* Cambridge, England: Cambridge University Press.

Levy, Joseph. 1976. *Un village de bout du monde: Modernisation et structure villageosie aux Antilles Francaise.* Montreal, Canada: Université de Montreal, Centre de Recherches Caraibes.

LiPuma, Edward, and Sarah Meltzoff. 1997. "The Cross-Currents of Ethnicity and Class in the Construction of Public Policy." *American Ethnologist* 24 (1): 114–31.

Lomnitz, Larissa. 1977. *Networks and Marginality.* New York: Academic Press.

Long, Norman. 1977. *An Introduction to the Sociology of Rural Development.* London: Tavistock.

Magnarella, Paul. 1979. *The Peasant Venture.* Cambridge, Mass.: Shenkman.

Mahler, Sarah. 1995. *American Dreaming: Immigrant Life on the Margins.* Princeton, N.J.: Princeton University Press.

Maiolo, John, and Michael Orbach, eds. 1981. *Modernization and Marine Fisheries Policy.* Ann Arbor, Mich.: Ann Arbor Press.

Maldonado, Marta María, Manuel Valdés Pizzini, and Alfonso R. Latoni. 2000. "Owning and Contesting El Yunque: Forest Resources, Politics, and Culture in Puerto Rico." *Berkeley Journal of Sociology* 44:82–100.

Maril, Robert Lee. 1995. *The Bay Shrimpers of Texas.* Lawrence: University Press of Kansas.

Martínez, José Eduardo, and Manuel Valdés Pizzini. 1996. "Culture, Rationality, and Development: Historical Constructions and 'Distortions' of Conservation Efforts in the Fisheries of Southwestern Puerto Rico." In *Proceedings of the Forty-ninth Gulf and Caribbean Fisheries Institute* (Barbados) 49:86–94.

Marx, Karl. 1967. *Capital.* Vol. 1. New York: International.

Massey, Douglas, Rafael Alarcón, Jorge Durand, and Humberto González. 1987. *Return to Aztlán: The Social Process of International Migration from Western Mexico.* Berkeley: University of California Press.

McCay, Bonnie. 1984. "The Pirates of Piscary: Ethnohistory of Illegal Fishing in New Jersey." *Ethnohistory* 31:17–37.

McCay, Bonnie, and James Acheson. 1987. *The Question of the Commons.* Tucson: University of Arizona Press.

McGoodwin, James R. 1990. *Crisis in the World's Fisheries.* Stanford, Calif.: Stanford University Press.

Meillasoux, Claude. 1972. *Economic Anthropology and Marxism.* Cambridge, England: Cambridge University Press.

Meléndez, Arturo. 1986. *La batalla de Vieques.* Río Piedras, Puerto Rico: Ediciones Huracán.

Meltzoff, Sarah. 1988. "The Crosscurrents of Ethnicity and Class: Conflict and Conservation in the Florida Keys." Paper presented at the Maritime Social Sciences Conference, University of South Alabama at Mobile, April.

Mintz, Sidney. 1956. "Cañemelar: The Subculture of a Rural Sugar Plantation Proletariat." In *The People of Puerto Rico,* by Julian Steward, Robert Manners, Sidney Mintz, Ellen Padilla, Raymond Scheele, and Eric Wolf. Urbana: University of Illinois Press.

———. 1960. *Worker in the Cane: A Puerto Rican Life History.* New York: W. W. Norton.

———. 1977. "The So-Called World-System: Local Initiative and Local Response." *Dialectical Anthropology* 2 (4): 253–70.

———. 1985. *Sweetness and Power.* New York: Viking.

Moberg, Mark. 1992. *Citrus, Strategy and Class.* Iowa City: University of Iowa Press.

Morris, Nancy. 1995. *Puerto Rico: Culture, Politics, Identity.* Westport, Conn.: Praeger.

Munro, John, and Ian Smith. 1983. "Management Strategies in Multi-species Complexes in Artisanal Fisheries." *Proceedings of the Annual Meetings of the Gulf and Caribbean Fisheries Institute* 36:120–132.

Naipaul, V. S. 1989. *A Turn in the South.* New York: W. W. Norton.

Nash, June. 1985. "Segmentation of the Work Process in the International Division of Labor." In *The Americas in the New International Division of Labor,* ed. Steven Sanderson. New York: Holmes and Meier.

———. 1994. "Global Integration and Subsistence Insecurity." *American Anthropologist* 96 (1): 7–30.

Nash, June, and M. Patricia Fernández-Kelley, eds. 1983. *Women, Men, and the New International Division of Labor.* Albany: State University of New York Press.

National Oceanographic and Atmospheric Association (NOAA). 1983. "Proposed La Parguera National Marine Sanctuary: Draft Environmental Impact Statement." Washington, D.C.: NOAA.

Newman, Katherine. 1988. *Falling from Grace: The Experience of Downward Mobility in the American Middle Class.* New York: Free Press.

North Carolina Division of Occupational Safety and Health. 1989. "OSHA Violation Report of Lewiston Perdue Plant Violations." Raleigh: North Carolina Department of Labor.

Nweihed, Kaldone. 1983. *El Caribe de la pesca.* Caracas, Venezuela: Asociación de Universidades e Institutos de Investigación del Caribe.

Ong, Aiwa. 1988. "The Production of Possession: Spirits and the Multinational Corporation in Malaysia." *American Ethnologist* 15 (1): 28–42.

Orlove, Benjamin. 1977. *Alpacas, Sheep, and Men.* New York: Academic Press.

Padilla, Elena. 1956. "Nocorá: The Subculture of Workers on a Government-Owned Sugar Plantation." In *The People of Puerto Rico,* by Julian Steward, Robert Manners, Sidney Mintz, Ellen Padilla, Raymond Scheele, and Eric Wolf. Urbana: University of Illinois Press.

Pálsson, Gisli, and E. Paul Durrenberger. 1983. "Icelandic Foremen and Skippers." *American Ethnologist* 10:511–28.

Paul, Benjamin, ed. 1955. *Health, Culture, and Community.* New York: Russell Sage.

Paz, Octavio. 1995. "In Search of the Present: 1990 Nobel Prize Acceptance Speech." *Georgia Review.* 49 (1): 255–264.

Pérez, Ricardo. 2000. "Fragments of Memory: The State and Small-Scale Fisheries Modernization in Southern Puerto Rico." Ph.D. diss., Department of Anthropology, University of Connecticut at Storrs.

Picó, Fernando. 1983. *La historia general de Puerto Rico.* Rio Piedras: University of Puerto Rico Press.

———. 1986. *Los gallos peleados.* Río Piedras: University of Puerto Rico Press.

Poggie, John. 1979. "Small-Scale Fishermen's Beliefs About Success and Development: A Puerto Rican Case." *Human Organization* 38:6–11.

———. 1980. "Small-Scale Fishermen's Psychocultural Characteristics and Cooperative Formation." *Anthropological Quarterly* 53:20–28.

Pollnac, Richard. 1981. "Sociocultural Aspects of Technological Change in Marine Fishing Communities." In *Modernization and Marine Fisheries Policy,* ed. John Maiolo and Michael Orbach, 225–47. Ann Arbor, Mich.: Ann Arbor Press.

Pollnac, Richard, and John Poggie. 1978. "Economic Gratification Orientations among Small-Scale Fishermen in Panama and Puerto Rico." *Human Organization* 37:355–67.

Pollnac, Richard, and J. Sutinen. 1979. "Economic, Social, and Cultural Aspects of Stock Assessment for Tropical Small-Scale Fisheries." In *Stock Assessment for Tropical Small-Scale Fisheries.* Kingston: University of Rhode Island, Center for Marine Resource Development.

Portes, Alejandro. 1997. "Immigration Theory for a New Century: Some Problems and Opportunities." *International Migration Review* 31 (4): 799–825.

Portes, Alejandro, and Robert Bach. 1985. *Latin Journey.* Berkeley: University of California Press.

Portes, Alejandro, and John Walton. 1979. *Labor, Class, and the International System.* New York: Academic Press.

Posada, Juan Manuel, Manuel Valdés Pizzini, Marcos Rosada, and Kurt Grove. 1996. "Mapping Fishing Grounds Using GPS Technology." *Proceedings of the Forty-ninth Gulf and Caribbean Fisheries Institute* (Barbados) 49:125–37.

Pratts, Saul J. 1987. *La política social en Puerto Rico.* Santurce, Puerto Rico: Jay-Ce Printing.

Price, Richard. 1966. "Caribbean Fishing and Fishermen." *American Anthropologist* 68:1363–83.

Rappaport, Roy. 1969. *Pigs for the Ancestors: Ritual in the Ecology of a New Guinea People.* New Haven, Conn.: Yale University Press.

Rebel, Hermann. 1989. "Cultural Hegemony and Class Experience: A Critical Reading of Recent Ethnological-Historical Approaches (Part Two)." *American Ethnologist* 16 (2): 350–65.

Redfield, Robert. 1947. "The Folk Society." *American Journal of Sociology* 52:293–308.

Richardson, Benjamin. 1983. *Caribbean Migrants.* Knoxville: University of Tennessee Press.

Robben, Antonius. 1989. *Sons of the Sea Goddess.* New York: Columbia University Press.

Rogers, Everett. 1969. *Modernizing Among Peasants: The Impact of Communication.* New York: Holt, Rinehart, and Winston.

Rosaldo, Renato. 1989. *Culture and Truth.* Boston: Beacon Press.

Roseberry, William. 1976. "Rent, Differentiation, and the Development of Inequality Among Peasants." *American Anthropologist* 78:243–65.

———. 1978. "Historical Materialism and *The People of Puerto Rico." Revista/Review Interamericana* 8 (1): 26–36.

———. 1983. *Coffee and Capitalism in the Venezuelan Andes.* Austin: University of Texas Press.

———. 1988. "Political Economy." *Annual Reviews of Anthropology* 17:161–85.

———. 1989. "Americanization in the Americas." In *Anthropologies and Histories: Essays in Political Economy.* New Brunswick, N.J.: Rutgers University Press.

Rouse, Roger. 1992. "Making Sense of Settlement." In *Towards a Transnational Perspective on Migration: Race, Class, Ethnicity and Nation Reconsidered,* ed. Nina Glick-Schiller, Linda Basch, and Cristina Szanton Blanc, 25–52. New York: Annals of the New York Academy of Sciences, No. 645.

Sahlins, Marshall. 1972. *Stone Age Economics.* Chicago: Aldine.

Sanderson, Steven, ed. 1985. *The Americas in the New International Division of Labor.* New York: Holmes and Meier.

Schultz, Theodore. 1964. *Transforming Traditional Agriculture.* New Haven, Conn.: Yale University Press.

Scott, James. 1976. *The Moral Economy of the Peasant.* New Haven, Conn.: Yale University Press.

———. 1985. *Weapons of the Weak: Everyday Forms of Peasant Resistance.* New Haven, Conn.: Yale University Press.

Sider, Gerald. 1986. *Culture and Class in Anthropology and History: A Newfoundland Illustration.* Cambridge, England: Cambridge University Press.

Smith, Carol. 1977. *Regional Analysis.* New York: Academic Press.

Smith, Suzanne, and Michael Jepson. 1993. "Big Fish, Little Fish." *Social Problems* 40 (1): 39–49.

Spicer, Edwin, ed. 1954. *Human Problems in Technological Change.* New York: Russell Sage Foundation.

Stavenhagen, Rodolfo. 1975. "Classes, Acculturation, and Internal Colonialism." In *Social Problems in Latin America,* ed. Irvin Horowitz. New York: Monthly Review Press.

Steward, Julian. 1938. *Basin Ethnography and Cultural Ecology.* Washington, D.C.: Bureau of American Ethnography.

———. 1955. *Theory of Cultural Change.* Urbana: University of Illinois Press.

Steward, Julian, Robert Manners, Sidney Mintz, Ellen Padilla, Raymond Scheele, and Eric Wolf. 1956. *The People of Puerto Rico.* Urbana: University of Illinois Press.

Stewart, Susan. 1997. "Our Ruin." *The Kenyon Review* 19 (1): 145–52.

Stichter, Sharon. 1985. *Migrant Labors: African Society Today.* Cambridge, England: Cambridge University Press.

Stoffle, Rich. 1986. *Caribbean Fishermen Farmers: A Social Assessment of Smithsonian King Crab Mariculture.* Ann Arbor, Mich.: Institute for Social Research.

Stoler, Ann. 1985. *Capitalism and Confrontation in Sumatra's Plantation Belt.* New Haven, Conn.: Yale University Press.

Stonich, Susan. 1998. *The Other Side of Paradise.* Boulder, Colo.: Westview Press.

Stull, Donald, Michael Broadway, and David Griffith, eds. 1995. *Any Way They Cut It: Meat Packing and Small-Town America.* Lawrence: University Press of Kansas.

Sued Badillo, Jalil, and Ángel López Cantos. 1986. *Puerto Rico Negro.* Río Piedras, Puerto Rico: Editorial Cultural.

Taller de Formación Política. 1982. *Huelga en la caña.* Río Piedras, Puerto Rico: Ediciones Huracán.

Tambiah, Sidney. 1987. "Ethnic Conflicts in the World Today." *American Ethnologist* 16:335–49.

Taussig, Michael. 1980. *The Devil and Commodity Fetishism in South America.* Chapel Hill: University of North Carolina Press.

———. 1987. *Shamanism, Colonialism, and the Wild Man: A Study in Terror and Healing.* Chicago: University of Chicago Press.

Thompson, E. P. 1974. *The Making of the English Working Class.* Cambridge, England: Cambridge University Press.

Tonnies, Ferdinand. 1955. *Community and Association.* London: Routledge and Kegan Paul.

U.S. Congress. 1978. "The British West Indies Temporary Alien Labor Program." Washington, D.C.: U.S. Government Printing Office.

U.S. Department of Labor. 1987. "Occupational Injuries and Illnesses by Industry." News of the Bureau of Labor Statistics (USDL-88-562). Washington, D.C.: U.S. Government Printing Office.

Valdés Pizzini, Manuel. 1985. "Social Relations of Production in Puerto de La Corona: Capitalism and Development in the Puerto Rican Fisheries." Ph.D. diss. State University of New York at Stony Brook.

———. 1987. "Apuntes sobre el desarrollo histórico de la pesca en Puerto Rico." University of Puerto Rico Sea Grant College Program Report (UPR-SG-28). Mayagüez: University of Puerto Rico.

———. 1989. "Quarterly Report to the Southeast Regional Office." St. Petersburg, Fla.: National Marine Fisheries Service/National Oceanographic and Atmospheric Administration.

———. 1990a. "Etnología crítica del trabajo en las pesquerías de Puerto Rico y el Caribe insular." *Caribbean Studies* 23 (1–2): 61–82.

———. 1990b. "Fishermen Associations in Puerto Rico: Praxis and Discourse in the Politics of Fishing." *Human Organization* 49 (2): 164–73.

———. 1995. "Fisheries Co-management in Puerto Rico." University of Puerto Rico Sea Grant College Program Progress Report. Mayagüez: University of Puerto Rico.

———. In press. "Dialogía y ruptura: La tradición etnográfica en la antropología aplicada en Puerto Rico, a partir de *The People of Puerto Rico.*" *Journal of Latin American Anthropology.*

Valdés Pizzini, Manuel, Alejandro Acosta, David Griffith, and Mervin Ruíz. 1996. "Assessment of the Socio-economic Impact of Fishery Management Options upon Gill Net and Trammel Net Fishermen in Puerto Rico: An Interdisciplinary Approach (Anthropology and Fisheries Biology) for the Evaluation of Management Alternatives." National Marine Fisheries Service Final Report. St. Petersburg, Fla.: National Marine Fisheries Service.

Valdés Pizzini, Manuel, Jaime Gutiérrez Sánchez, and Ruperto Chaparro. 1988. "An Inventory of Recreational Fishing Infrastructure in Puerto Rico." National Marine Fisheries Service Technical Report. St. Petersburg, Fla.: National Marine Fisheries Service/National Oceanographic and Atmospheric Administration, Southeast Regional Office.

Vayda, Andrew. 1979. *Environment and Cultural Behavior.* Garden City, N.Y.: Natural History Press.

Vélez, Martín, S. S. Díaz Pacheco, and P. R. Vázquez Calcerrada. 1945. *La pesca y distribución de pescado en Puerto Rico.* Río Piedras: University of Puerto Rico Agricultural Experiment Station.

Vickers, David. 1994. *Farmers and Fishermen: A History of Essay County, Massachusetts.* Chapel Hill: University of North Carolina Press.

Wallerstein, Immanuel. 1974. *The Modern World System.* New York: Academic Press.

Wessman, James. 1977. "Towards a Marxist Demography: A Comparison of Puerto Rican Landowners, Peasants, and Rural Proletarians." *Dialectical Anthropology* 1 (4): 223–33.

Wieler, D. M., and J. R. Suárez-Caabro. 1980. "Perspectiva de las estadísticas de la pesca en pequeña escala de Puerto Rico." CODREMAR Informe Técnico 1 (1): 1–27.

Williams, Raymond. 1977. *Marxism and Literature.* Oxford, England: Oxford University Press.

Wolf, Eric. 1966. *Peasants.* Englewood Cliffs, N.J.: Prentice-Hall.

———. 1982. *Europe and the People Without History.* Berkeley: University of California Press.

———. 1999. *Envisioning Power.* Berkeley: University of California Press.

Worsley, Peter. 1984. *The Three Worlds.* Chicago: University of Chicago Press.

Wulff, Robert, and Shirley Fiske. 1987. *Anthropological Praxis: Translating Knowledge into Action.* Boulder, Colo.: Westview Press.

# Index